Modern NMR Spectroscopy

A Workbook of Chemical Problems

Second Edition

Modern NMR Spectroscopy

A Workbook of Chemical Problems

Second Edition

JEREMY K. M. SANDERS and
EDWIN C. CONSTABLE

University Chemical Laboratory
University of Cambridge, UK

BRIAN K. HUNTER

Department of Chemistry,
Queen's University at Kingston, Ontario, Canada

and

CLIVE M. PEARCE

Schering Agrochemicals Ltd, UK

Oxford   New York   Toronto
OXFORD   UNIVERSITY   PRESS
1993

Oxford University Press, Walton Street, Oxford OX2 6DP
Oxford New York Toronto
Delhi Bombay Calcutta Madras Karachi
Kuala Lumpur Singapore Hong Kong Tokyo
Nairobi Dar es Salaam Cape Town
Melbourne Auckland Madrid
and associated companies in
Berlin Ibadan

Oxford is a trade mark of Oxford University Press

Published in the United States
by Oxford University Press Inc., New York

© Jeremy K. M. Sanders, Edwin C. Constable, Brian K. Hunter, and Clive M. Pearce
1989, 1993

First edition published 1989
Reprinted 1990 (with corrections), 1992
Second edition published 1993

A catalogue record for this book is available from the British Library

Library of Congress Cataloging in Publication Data
Sanders, Jeremy K. M.
  Modern NMR spectroscopy: a workbook of chemical problems /
Jeremy K. M. Sanders and Edwin C. Constable: Brian K. Hunter and
Clive M. Pearce.—2nd ed.   Includes index.
  1. Nuclear magnetic resonance spectroscopy—Problems, exercises,
etc.   I. Constable, Edwin C.   II. Title.
QD96.N8S25   1993   543′.0877—dc20   93-10834
ISBN 0 19 855812 0

Typeset by the authors
Printed in Great Britain on acid-free paper by the Alden Press, Oxford

Preface

One of the important skills that a practical chemist needs is the ability to interpret NMR spectra. In this workbook we aim to develop that skill to an advanced level by a combination of worked examples and set problems covering one- and two-dimensional techniques applied to organic and inorganic systems. We also hope to illustrate the exquisite chemical insight that can be obtained from careful design of spectroscopic experiments and from detailed consideration of the resulting spectra. The book is designed to be used by classes or individuals who have our main text, *Modern NMR spectroscopy - a guide for chemists*. It can also be viewed as a resource for the teacher who wants to show examples of specific NMR effects or who needs a change from the tired set problems of the past.

The simulated multiplets on p. 9 were produced for us by Jeremy Titman. All the other multiplets and spectra (even the traces in problem 3.9) are authentic. Samples or spectra were donated by Stuart Amor, John Anderson, Mary Baum, Beat Ernst, Richard Ernst, Malcolm Green, Richard Hibbert, Chris Hunter, John Kennedy, Tony Kirby, Antonin Lycka, Gavin McInnes, Stephen Matlin, R. Mynott, Clive Pearce, Bill Smith, Tammo Winkler, Victor Wray and Herman Ziffer.

Sources of published spectra or results are given in the Solutions chapter. Permission to reproduce these spectra has been granted by Academic Press, the American Chemical Society, Elsevier Sequoia S.A., the National Research Council of Canada, the Royal Society of Chemistry and John Wiley.

Where no reference or acknowledgement is given, the spectrum has not been previously published. Some of these spectra derive from work in our laboratories over many years, while many were run especially for this book by Clive Pearce; a few were run for us by Brian Crysell, Mike Kinns, and Steve Wilkinson. The remainder were donated by the friends and colleagues mentioned above. In several problems, we have 'adjusted' the structures or results to make them more accessible, while a few of the problems have been so modified by years of class use in Cambridge that the original literature sources and structural details have been lost. We apologize to any authors whose work has suffered in this way, and hope that they will let us know so that we can acknowledge them in any future editions.

Generations of Cambridge undergraduates have tested many of the older problems. Harry Anderson, Richard Bonar-Law, Tony Kirby, Clive Pearce and David Ryan have tested and proof-read some or all of the more recent problems and have made useful suggestions for improving them.

The staff of Oxford University Press have given us invaluable design support, and Jean Jacobs helped with emergency typing. Louise Sanders has again acted as our English expert, and Mike Springett was responsible for lettering all spectra and for the final paste-up. All the remaining errors are, of course, our responsibility.

Once again, our families and students have suffered more, and for longer, than they should have.

To all those mentioned above, we are most grateful.

Cambridge and   J.K.M.S.
Kingston   E.C.C.
May 1989   B.K.H.

Preface to the second edition

This revised and enlarged edition contains seven new problems (2.12 and 3.15–3.20). Some of these problems are derived from work published since 1989 and some from our own observations. New spectra are reproduced with permission of the American Chemical Society, the Royal Society of Chemistry, and Neue Schweizerische Chemische Gesellschaft. We have also updated the cross references to our main book to be consistent with the recently published second edition, and have corrected the absolute stereochemistry of camphor. We are grateful to Debbie Crans, Guella Graziano and Brian Mann for copies of their spectra, and Mike Springett for more lettering and pasting.

Cambridge and   J.K.M.S.
Kingston   B.K.H.
February 1993   C.M.P.
                                                  E.C.C.

Contents

Introduction and guide

Chemists use a wide range of physical techniques for studying the structures and reactions of the molecules they are interested in. One of the skills they need is the ability to choose and exploit the most appropriate technique for studying the particular molecules of interest. X-ray crystallography is the ultimate arbiter of chemical structure and in many cases it is now the method of choice; the use of automated data collection and direct methods of structure solution have reduced many problems to a routine level. However, crystallography has many limitations beyond the obvious need for crystals: it cannot tell us anything about solutions, however pure they may be, or conformational equilibria, or complex mixtures or reaction kinetics. For this type of information, the chemist must turn to other physical techniques, such as some form of spectroscopy.

NMR is undoubtedly the most versatile of all the spectroscopic techniques because it embraces the widest range of materials: liquids and solids, organic and inorganic molecules, pure compounds and complex reaction mixtures can all be studied, dissected and understood by carefully designed NMR experiments. NMR techniques have advanced dramatically in the past few years and are now more powerful than ever before. The theory and practice of many of the new NMR experiments available to the chemist have been described in a number of recent textbooks, which range in character from mainly descriptive to highly mathematical.[1-5] Several books at the general undergraduate level also describe some of the new techniques.[6-9] However, knowing how an experiment works and being able to exploit it efficiently are quite different skills. This workbook is about this second skill.

Aims and scope

The aim of this workbook is to give you experience in interpreting the appearance of NMR spectra, and in applying those interpretations in a chemical context. You will then be able to approach real spectroscopic and chemical problems with confidence. The emphasis is on practical applications rather than theory, so there are no problems testing your understanding of how pulse sequences work.

We have designed this workbook to be used in conjunction with our main book, *Modern NMR spectroscopy – a guide for chemists*,[1] and have given many cross-references to help you. The cross-references to our main book are in the form §$x.y$, which means Chapter x, Section y. Several other textbooks[2-8] have much in common with ours and they might be useful alternatives to it. This workbook is not a compilation of chemical shifts and coupling constants. You will probably need one of these too.[6,8,10,11] However, at the end of this Introduction we have provided a small amount of background material, a list of the conventions and abbreviations we have used and a table listing the relevant properties of all of the NMR-active isotopes mentioned in the problems.

Most of the chapters contain both worked examples and problems for you to solve yourself. As you progress through a chapter, problems tend to become more complex or difficult but this is not an absolute rule. They cover a range of practical aspects, from the appearance of badly run or wrongly processed spectra, through spectrum assignment to the determination of reaction mechanisms and complex structures. Organic and inorganic examples are interspersed, as are one- and two-dimensional spectra. Spectra of several metal nuclei are included along with the historically more conventional nuclei. We hope that there is a large enough number of each type of problem to satisfy organic or inorganic specialists, but believe that modern chemists should be able to tackle the whole range of problems.

The worked examples, which range from simple to very advanced, are clearly marked except in Chapter 1, which consists entirely of worked examples. The examples are designed to give you guidance on how to approach different types of problem. Each logical step in the analysis of the spectrum is given in detail so that you can see exactly where each conclusion comes from. This can make the worked examples look rather long, but the effort of working through them is well worth while. The advanced examples are intended to illustrate just how much chemical detail can be extracted from spectra by careful thought and analysis. Some of them are actually more difficult than any of the problems.

We have assumed that, in addition to knowing something about chemical shifts and *J*-coupling, you also have at least an undergraduate knowledge of the rules of valency and bonding, geometry and stereochemistry, and some simple chemical properties such as basicity. The material in Chapter 5 also requires a working knowledge of other spectroscopic methods and chemical reactions.

Several of the spectra in this workbook were obtained using techniques such as proton-detected C–H shift correlation and multiple-quantum-filtered phase-sensitive COSY which were not covered in detail in our main text because they have come into general use since it was written. This is a measure of the rate at which practical NMR is progressing but presents no problem in the interpretation of the spectra. Various field strengths and modes of presentation, ranging from continuous-wave traces to phase-sensitive two-dimensional contour plots, were used for the spectra. In part, this reflects the history of individual problems, but it is also intentional. It is important to be able to extract the essential message of a spectrum independently of the way it is presented.

Chapter 1 focusses on the setting up and processing of spectra. It consists of a set of spectra that are wrongly processed or acquired, and its title reflects the experience of almost everyone who has tried to run Fourier transform (FT) NMR spectra. All the spectra are of the same sample of camphor, a convenient compound that should be readily available to anyone who wishes to repeat the experiments for themselves. If you do not run your own spectra, you should still find it instructive to work through these examples as they illustrate some of the dangers of simply accepting the appearance of a spectrum that has been run for you.

We assume that the spectrometer is working well and is properly shimmed, and concentrate on the two most common types of error committed by the novice. These are (i) collecting a good-quality free induction decay (FID) but processing it wrongly and (ii) collecting an inappropriate FID because the wrong values are used for parameters such as spectral width, pulse width or decoupler settings. The chapter begins with a series of ^{1}H spectra derived from a single FID; the spectra differ only in the way that the FID has been processed. These are followed by a series of ^{1}H spectra illustrating some errors of the second kind. We then switch to ^{13}C spectra and examine the problem of acquiring good spectra when ^{1}H decoupling or complex pulse sequences are required. The final part of the chapter examines of some of the problems which can be introduced into two-dimensional spectra through operator error. The chapter consists entirely of worked examples with brief answers immediately following the questions. Cross-references are given to the more detailed explanations that can be found in Chapter 1 of our main text.

Chapter 2 is concerned with the interpretation of spectra. It tests your ability to assign and interpret the spectra of compounds with known structures and explores how the appearance of a spectrum depends on factors such as nuclear spin, isotopic abundance, molecular tumbling, scalar coupling, symmetry, conformation and stereochemistry.

A general feel for the sizes of chemical shifts and coupling constants will be helpful for the organic problems in Chapters 2–5, but we do not place much faith in detailed shift predictions based on correlation tables: the chemical literature abounds with 'confident' assignments that are based on chemical shift arguments but turn out to be wrong. In many of the problems, the sizes of coupling constants are more important than the chemical shifts. For example, it helps to know that J_{BC} in pyridine, **a**, has a standard aromatic value of 8 Hz, but that the electronegative nitrogen reduces J_{AB} to 5 Hz. Similarly, geminal (two-bond) proton–proton couplings are increased to 14–18 Hz by adjacent π-systems, but are reduced by electronegative substituents; in **b**, $J_{DE} = 16$ Hz while $J_{FG} = 8.5$ Hz.

Coupling constants in inorganic systems are less predictable apart from one-bond couplings, which are always very large. Two-bond couplings in an X–Y–Z fragment are much larger when the bond angle at Y is 180° than when it is 90°. This is a useful diagnostic probe of stereochemical relationships in octahedral complexes.

More than any other spectroscopic technique, NMR allows us to determine the symmetry of a molecule, and observe the often-unexpected chemistry that is occurring in a solution. These symmetry and exchange effects are explored in Chapter 3.

Chapter 4 tests your ability to determine the structures of molecules. The only information provided is the molecular composition and selected NMR properties, but it is possible to solve virtually all the structures unambiguously; where this is not the case, the problem clearly says so. We discuss below an outline strategy for solving structures using spectroscopic data.

Chapter 5 is similar to Chapter 4, but also includes other types of spectroscopic information and details of the chemical reactions that lead to the structures. Some of these problems will require an advanced undergraduate knowledge of chemistry; solving the problem will not only give you experience in interpreting spectra, but may also give you new insight into the mechanisms of chemical and biological processes.

If you cannot see how to approach or solve any particular problem, then you can turn to the Hints section. This contains clues for many, but not all, of the problems. In some cases the hints may suggest a starting point for your analysis and in others it will ask a further question to point you in the right direction. Some of the hints simply give a reference to an appropriate section of our main book.

Finally, there is a section called Solutions. In some cases, you will find a completely assigned spectrum, the correct structure or a detailed explanation, while in others there is only a reference to the original literature.

How to solve structures

There is no completely rational or reliable strategy for solving structures from spectroscopic data. The precise route to a structure will depend on how much background and spectroscopic information is available about the molecule, and on which of the spectra appear most useful. Nevertheless, there are some useful rules and guidelines about the right way to tackle such a problem. Further details will be found in Chapter 5 of our main book.

For an organic compound the first step is usually to find the molecular formula, probably from the mass spectrum, and to calculate the number of double bond equivalents (DBEs). An acyclic saturated hydrocarbon has the formula $C_N H_M$ where $M = 2N+2$. Each double bond or ring in the molecule reduces the value of M by two. So if $M = 2N$ the molecule has one DBE; we cannot tell from the formula whether it is in the form of a ring or unsaturation. A benzene ring corresponds to 4 DBEs: three double bonds and a ring. The presence of oxygen or other divalent elements does not affect the value of M. Each monovalent atom such as chlorine can be treated as a proton for the purpose of calculation, while one proton has to be subtracted for each trivalent atom such as nitrogen.

If infrared and ultraviolet spectra are available, they should be inspected for preliminary clues about the functional groups and conjugation that might be present. A fairly superficial survey of the ^{1}H and ^{13}C spectra should enable you to assess such factors as the degree and kind of symmetry in the molecule; the ratio of aromatic to aliphatic carbons; the number of methyl, methylene, methine and quaternary carbons; and the number of exchangeable protons. The next stage is a detailed study of the one-dimensional or COSY ^{1}H spectrum:

Look for a good 'starting point', such as a methyl group, a slowly-exchanging hydroxyl proton or an obviously aromatic signal.

A good starting point will usually have a characteristic chemical shift or other feature that makes it easy to assign.

Find all the signals that are connected to your starting point by *J*-coupling, then find all the signals that are connected to those by *J*-coupling, and so on. Continue until you come to the end of that molecular fragment and cannot go any further.

Find a fresh starting point and repeat the search for *J*-connections. Continue with succeeding starting points until as much of the spectrum as possible is collected up into a series of fragments.

At this stage you have pieces that need to be fitted together like a jigsaw puzzle. Try to piece the fragments together using NOE connections or, if you have them available, long-range C–H correlations. If the structure is at all complicated, you will realize that the spectra have not given you the answer. Spectra and chemical reactions alone can never tell you the structure of a compound. All they can do is to give you pieces of information.

Now you have to be creative and invent a series of structures that appear to satisfy all the spectroscopic criteria you have so far, and test each structure critically against the spectra. Predict the chemical shifts; do they really match? Predict all the multiplet patterns; do they match well? Predict the NOEs and mass spectral fragmentation patterns and compare them with the experimental results. Discard any structures that clearly fail a test. Those that pass should be subjected to further, more searching tests.

With luck, you will now have only one structure left. Often several candidates fit all the evidence you have so far, and you will have to devise a new experiment to distinguish the possibilities. On other occasions you will find that you have discarded all your candidate structures and will have to think of new structures for testing. Ultimately, you should have only one convincing structure.

Inventing new candidate structures is surprisingly difficult because we tend to think only of structural types that we already know well. Many complex natural product structures have supposedly been elucidated in recent years but a significant proportion of them are wrong. The proposed structures are all apparently compatible with the available spectroscopic data and are usually reasonable. But they are by no means the only structures that are compatible with the data. Usually the authors had failed to think of the correct structure even as a possibility. In most cases we still do not know which are right and which are are wrong. Errors are only discovered when different research groups work independently on the same compound, or when a synthesis is attempted or when a structure is firmly established by X-ray crystallography. Taylor has given an excellent description of a recent example of how a structure was 'corrected' several times before the real structure was elucidated.[12]

Clearly the fundamental problem is the invention of new structures. There is a great need for reliable computer algorithms that can generate structures and test them against a set of spectroscopic constraints. Some programs are available,[13] but at present there is no real substitute for the intuition and hard work of the chemist. None of the problems in this workbook are so difficult that they need computer assistance.

It is important to realize the difference between the determination of structure and the determination of conformation. There is only one correct structure for a molecule. Either a proposed structure is right or it is wrong, and a well-designed set of experiments should clearly distinguish right from wrong. If a compound is pure, all the molecules will have identical structures. This picture is slightly complicated by compounds that exist in different tautomeric forms but all the molecules of one form are again identical.

The picture is quite different for conformations. A molecule such as *t*-butyl cyclohexane is generally said to exist in a single conformation, with an equatorial *t*-butyl group, but it has a significant population of chairs with axial *t*-butyl groups. Worse, the 'chair' conformation sits at the bottom of a relatively shallow potential well, so we have an ensemble of molecules with a range of very similar, but not identical, conformations. The internuclear distances in these similar conformations differ and the resulting NOEs between nuclei are complex time-averages from the different conformations. The inverse sixth-power dependence means that the observed NOE are heavily weighted towards those conformations that allow the closest approach of the nuclei that relax each other. The implications for the determination of conformation are severe; for details see §6.5 and elsewhere.[14] In this workbook, questions about conformation are about the dominant conformation and are mostly qualitative. We can almost never rule out the presence of minor conformations and we can sometimes detect them. In *very* favourable cases, such as problem E3.3 where there are two well-defined possible conformations, we can even use NOEs to estimate the equilibrium constant between the two conformers.

Analysis of multiplets

Nuclei with $I = 1/2$

Multiplets containing several different couplings can be surprising and complex in appearance but they are often easier to analyse than they seem. Each coupling splits the resonance into two lines, so in principle if a nucleus has N couplings to spin-$1/2$ nuclei, its resonance signal will have 2^N lines. However, when two or more couplings are equal, or there is some other fortuitous relationship between them, fewer lines will be resolved. A triplet is then merely a doubled doublet ('dd') where two equal splittings lead to a coincident central line. Some common multiplets are shown opposite to illustrate these points. The multiplets on the left- and righthand sides differ only in the size of a single coupling.

There are two useful tricks for analysing multiplets. These are also apparent in the illustrations:

> The distance, in Hz, between the outermost two lines of a multiplet is equal to the sum of all the couplings in the multiplet.

> The couplings can often be found by measuring the separations of successive inner lines from one outer line.

If the separation between two resonances is Δ Hz and their coupling is J Hz, then the rules above hold if $\Delta \gg J$. The multiplet is said to be 'first-order'. When $\Delta/J < 5$, distortions and extra lines appear, and the multiplet is then 'second-order'. See Abraham *et al* [3] for a good description of how to analyse such spectra. A first-order analysis is adequate for solving all the problems in this workbook, although many of the multiplets are distinctly second-order.

Nuclei with $I > 1/2$

These nuclei often relax too rapidly for J-coupling to be seen (§1.4.10 and §7.2.3) but coupling is occasionally resolved. There are several examples in this workbook of coupling to spins with $I > 1/2$. In such cases the rule governing splitting patterns is a simple extension of the spin-$1/2$ case: for each coupling to a nucleus with spin I, a resonance is split into $2I + 1$ lines of equal intensity. So, a proton or a carbon coupled to a single deuteron ($I = 1$) is a 1:1:1 triplet. There is an example of this type of multiplet in Chapter 2. It is also a familiar sight to anyone who has looked at the 77 ppm region of a ^{13}C spectrum run in $CDCl_3$ solution.

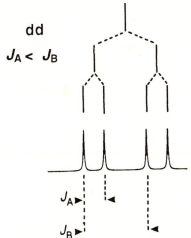

dd

$J_A < J_B$

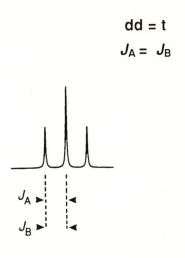

dd = t

$J_A = J_B$

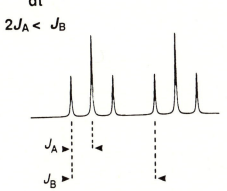

dt

$2J_A < J_B$

dt

$2J_A = J_B$

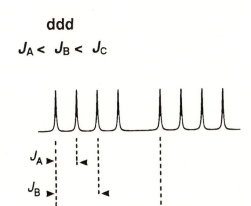

ddd

$J_A < J_B < J_C$

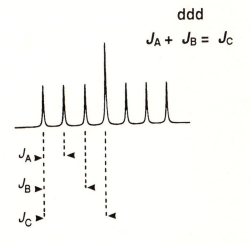

ddd

$J_A + J_B = J_C$

We can calculate the appearance of a proton coupled to two deuterons (as in a CHD_2 group) because each of the lines of the 1:1:1 triplet is split into a new 1:1:1 triplet. This can be easily represented as an arithmetical sum, where the top line is the left-most line of the triplet:

	1	1	1		
		1	1	1	
			1	1	1
Total	1	2	3	2	1

The predicted 1:2:3:2:1 quintet is precisely what we observe for the residual proton resonance in deuterated solvents such as acetone-d_6, DMSO-d_6 or methanol-d_4.

The effect of a third deuteron is calculated simply by splitting each of the quintet lines into three further lines of equal intensity and then summing:

	1	2	3	2	1		
		1	2	3	2	1	
			1	2	3	2	1
Total	1	3	6	7	6	3	1

This multiplet is seen in the carbon spectra of $^{13}CD_3$ groups.

It is straightforward to extend and apply these rules to resonances that are coupled to spins with $I > 1$ and to resonances that are coupled to several spins with different I-values.

Conventions and abbreviations

We have used the following conventions and abbreviations throughout:

NMR

SMALL CAPITALS are used for resonances whose assignments are unknown.

Chemical shifts for ^{13}C and ^1H are given in δ units, i.e. ppm from TMS. Where it is obvious that chemical shifts are being listed, the units may be missing. Integrals for resonances are indicated as 1X, 2X, etc., where X may be protons, carbons or fluorines.

J-couplings are in Hz if no units are given. Leading superscripts indicate the number of bonds connecting the coupled nuclei; trailing subscripts list the nuclei involved in the coupling, so $^2J_{PC}$ means coupling between phosphorus and carbon separated by two bonds, $P–X–^{13}C$.

The appearance of signals is described by the codes in the table below. In ^{13}C NMR spectra this means the appearance of the carbon signal acquired without decoupling and under low resolution so that long-range C–H couplings can be ignored.

Code	^1H	^{13}C
s	singlet	quaternary carbon without attached protons
d	1:1 doublet	methine (CH) carbon
t	1:2:1 triplet	methylene (CH_2) carbon
q	1:3:3:1 quartet	methyl (CH_3) carbon
qn	1:4:6:4:1 quintet	
m	complex multiplet	
br s	broad singlet	

In the examples and problems, resonances are described in a standard compact format. For example,

A   5.7   2H   ddt   12, 7, 9

means that resonance A has a chemical shift of 5.7 ppm and has an integral corresponding to two protons with a doublet splitting of 12 Hz, a further doublet splitting of 7 Hz and a triplet splitting of 9 Hz. Note that a multiplet with two different couplings has four equally intense lines and is a 'dd', NOT a quartet.

Exchangeable protons, such as OH, NH or SH, are marked * on spectra or in lists. { } means that the resonance inside the brackets is irradiated in a decoupling or pre-irradiation experiment.

In tables showing the outcome of NOE or decoupling experiments, the irradiated protons are listed vertically and the resonances affected by irradiation are listed horizontally. Enhancements are positive unless otherwise indicated.

An 'AB quartet' is a pair of *J*-coupled signals that are close in chemical shift. The inner lines of each doublet are more intense than the outer lines.

Mass spectrometry[6,8,9,15]

All the mass spectra listed are of positive ions.

m/z   Integral masses of significant ions. In every case in this book, the charge, z, is 1.

EI   Spectrum obtained with electron impact ionization.

CI   Spectrum obtained with chemical ionization.

FAB Spectrum obtained with fast atom bombardment ionization.

M^+ Indicates that the ion is the intact molecular ion.

MH^+ Indicates that the ion is a protonated intact molecular ion with mass M+1.

<u>105</u> Ions that are structurally most helpful are underlined.

Miscellaneous

Temperatures are quoted in K.

In structural diagrams of organic molecules we use the convention that hydrogen atoms are usually implied, i.e. carbon atoms with only two bonds drawn also have two attached protons. The main exceptions are cases where stereochemistry is important and for groups such as OH, NH and SH. A wavy line may mean one unknown stereochemistry or it may mean a mixture of stereochemistries. This will usually be clear from the context of the question.

ν Frequency of significant infrared absorptions in cm^{-1}.

λ Wavelength of maximum ultraviolet absorption in nm.

MW means integral molecular weight based on H = 1, C = 12, N = 14, O = 16, F = 19, S = 32.

Table of isotopes

The table below lists all the NMR-active nuclei that appear in Chapters 1–5, together with all their relevant properties. For a more complete listing, see Brevard and Granger.[16] Some nuclei, such as ^{18}O, have $I = 0$ and so are not NMR-active. However, they can sometimes be detected through the isotope shift that they induce in the resonances of nearby nuclei.

Nucleus	I	Nat. abundance (%)	Rel. frequency (MHz)
^{1}H	1/2	99.98	100.00
^{2}H	1	0.016	15.35
^{6}Li	1	7.42	14.71
^{10}B	3	19.58	10.75
^{11}B	3/2	80.42	32.08
^{13}C	1/2	1.11	25.14
^{14}N	1	99.63	7.22
^{15}N	1/2	0.37	10.13
^{19}F	1/2	100.0	94.08
^{31}P	1/2	100.0	40.48
^{47}Ti	5/2	7.28	5.637
^{49}Ti	7/2	5.51	5.638
^{51}V	7/2	99.76	26.29
^{119}Sn	1/2	8.58	37.27
^{181}Ta	7/2	99.98	11.97
^{187}Os	1/2	1.64	2.30
^{195}Pt	1/2	33.8	21.50
^{199}Hg	1/2	16.84	17.83

References

1.   Sanders, J.K.M. and Hunter, B.K. (1993). *Modern NMR spectroscopy – a guide for chemists*, 2nd edn, OUP, Oxford.

2.   Derome, A.E. (1987). *NMR techniques for chemical research*, Pergamon, Oxford.

3.   Abraham, R.J., Fisher, J. and Loftus, P. (1988). *Introduction to NMR spectroscopy*, Wiley, New York and London.

4.   Freeman, R. (1988). *A handbook of nuclear magnetic resonance*, Longman, Harlow.

5.   Ernst, R.R., Bodenhausen, G. and Wokaun, A. (1986). *Principles of nuclear magnetic resonance in one and two dimensions*, Clarendon Press, Oxford.

6.   Williams, D.H. and Fleming, I. (1987). *Spectroscopic methods in organic chemistry,* 4th edn, McGraw-Hill, Maidenhead and New York.

7.   Ebsworth, E.A.V., Rankin, D.W.H. and Cradock, S. (1991). *Structural methods in inorganic chemistry,* 2nd edn, Blackwell Scientific Publications, Oxford.

8.   Brown, D.W., Floyd, A.J. and Sainsbury, M. (1988). *Organic spectroscopy,* Wiley, New York and London.

9.   Sternhell, S. and Kalman, J.R. (1987). *Organic structures from spectra,* Wiley, New York and London.

10.   Pretsch, E., Seibl, J., Simon, W. and Clerc, T. (1991). *Tables of spectral data for structure determination of organic compounds* (trans. K. Biemann), 2nd edn, Springer-Verlag, Berlin, Heidelberg and New York.

11.   Breitmaier, E. and Voelter, W. (1987). *Carbon-13 NMR spectroscopy,* 3rd edn, VCH, Weinheim and New York.

12.   Taylor, D.A.H. (1986). *Tetrahedron*, **43**, 2779–87.

13.   Gray, N.A.B. (1986). *Computer-assisted structure elucidation,* Wiley, New York and London.

14.   Neuhaus, D. and Williamson, M.P. (1989). *The nuclear Overhauser effect in structural and conformational analysis,* Verlag Chemie, New York.

15.   Howe, I., Williams, D.H. and Bowen, R.D. (1981). *Mass spectrometry – principles and applications,* McGraw-Hill, Maidenhead and New York.

16.   Brevard, C. and Granger, P. (1981). *Handbook of multinuclear NMR,* Wiley, New York and London.

1 Errors I have made

In this chapter we explore some of the problems that you may encounter when running what are meant to be routine spectra. The increasing computer control found in modern spectrometers is a great benefit to the user, but can also lead to some subtle types of operator error. The problems discussed in this chapter are not the sort one encounters as a result of a failure in the system. Many of them appear when a parameter is wrongly set through forgetfulness or ignorance, or when a command is mistyped or misused.

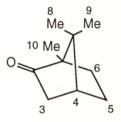

All the spectra in this chapter were obtained from a $CDCl_3$ solution of camphor run on a 9.4 tesla spectrometer (400 MHz for ^1H; 100 MHz for ^{13}C). To keep experiment times short, the solution was rather concentrated, but it was carefully prepared: the solution was filtered into a new, high-quality, 5 mm NMR tube and was degassed by bubbling oxygen-free nitrogen through the solution for several minutes. The tube was sealed with a tight-fitting cap. The molecule is tumbling rapidly, so the spins are relaxing slowly and the linewidths are intrinsically very small. The magnet was well shimmed for all the spectra, so any problems in linewidth are entirely due to errors in operation of the spectrometer.

1.1   We start with a good 400 MHz ^1H spectrum of the camphor solution described above. For this FID, 32K time domain points (=16K complex points) were acquired using a spectral width of 1600 Hz, giving an acquisition time of 10.2 s. This long acquisition time is necessary if one is seeking to obtain the very high resolution spectra which can, in principle, be obtained from this particular sample. Quadrature detection was used, with sequential acquisition of the two quadrature signals (§1.4.2). Eight transients were collected using $\pi/4$ pulses and a 5 s relaxation delay. In this sample, the spin-lattice relaxation is very slow and long recycle times (15 s) are necessary to avoid errors in signal intensities (§1.4.7).

Good quality spectra from this FID are shown on the next page. The spectra in problems 1.2–1.6 were all generated from the same free induction decay (FID) using data processing errors that are common to both experienced and inexperienced operators.

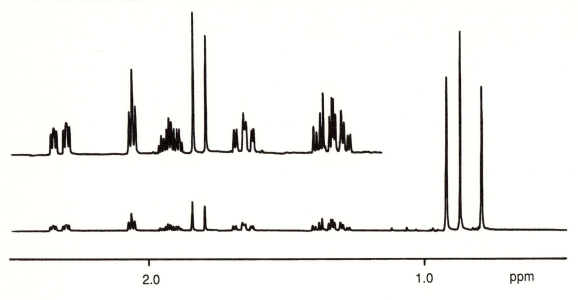

The spectrum above is displayed from 2.5 to 0.5 ppm; the only signal not shown is that from residual CHCl$_3$ in the solvent. Part of the spectrum is shown at higher vertical gain. For the purposes of this chapter it does not matter which signal belongs to which proton, but see §5.2.1 and §6.5.1 for how the spectrum was assigned using COSY and NOEs. The spectrum below shows the 2.0–1.75 ppm region of the same spectrum. Problems that concentrate on the detailed appearance of the spectrum use either this expansion or show the three methyl singlets from 0.8 to 1.0 ppm.

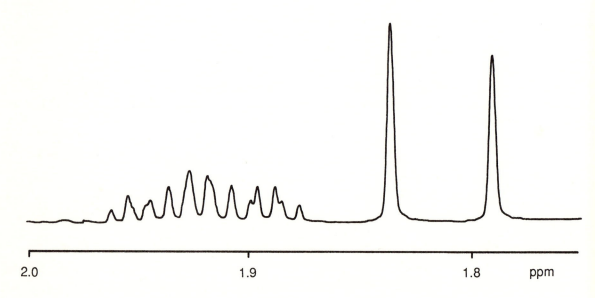

1.2 This partial spectrum was obtained from the same FID. Which error in data processing would cause the loss of resolution seen here? The resolution can be restored by changing a single parameter and reprocessing the data.

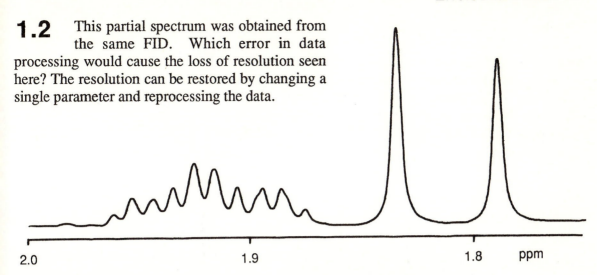

Answer The data were treated with an exponential multiplication before Fourier transformation to improve the signal-to-noise ratio in the frequency spectrum (§1.3.4). The 'line broadening' parameter used was 1 Hz, forcing a rapid decay of the FID. This gives a greater linewidth than is imposed by the field homogeneity and relaxation rates, and the apparent resolution is degraded. The digital resolution is 0.1 Hz, so a line broadening of 0.2 Hz could be used to reduce noise with essentially no effect on the apparent resolution (§1.3.4).

1.3 What data-processing routine has caused the distortion of the lineshapes below? Again the same FID has been used.

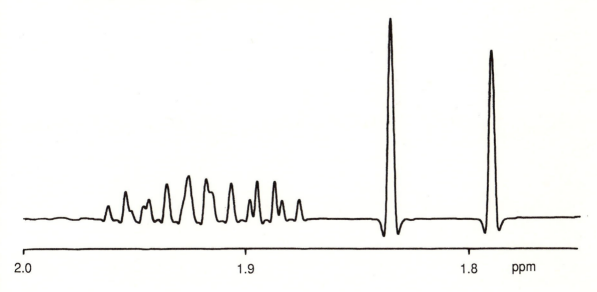

Answer Gaussian multiplication has been applied to the FID to improve the resolution. The line broadening (LB) was set to -1.0 Hz and the Gaussian maximum (GB) to 0.1. The resulting spectra have distorted lineshapes and intensities. If we attempt to enhance the resolution still further using a GB of 0.15 and an LB of -2.0 Hz, the spectrum becomes almost unrecognizable, as shown below. These parameters must be optimized for each spectrum, or even each signal.

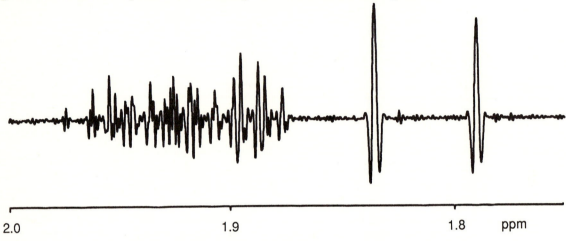

1.4 In the example below a negative exponential multiplication was applied to the FID before Fourier transformation. Why does this cause degradation of the spectrum?

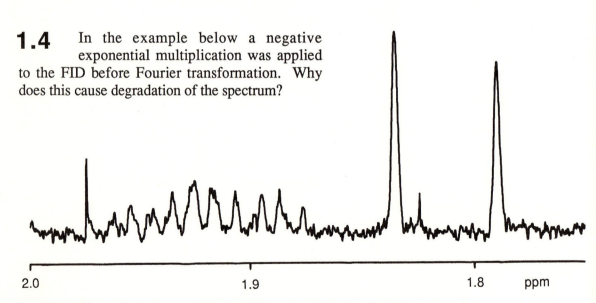

Answer First, the later end of the FID, containing *relatively* more noise, is being emphasized, producing a noisier spectrum. Second, the FID now contains a step at the end of the acquisition, giving rise to the spikes, **S**, which appear in the spectrum. With a more negative value for the line broadening, the FID will become so distorted that, after Fourier transformation, the result can be unrecognizable as a spectrum. It is very easy to make this mistake as the line broadening applied during a Gaussian multiplication is negative and the computer uses the same parameter for exponential or Gaussian multiplications. In making it easy to use 'old' parameters, the software enhances this kind of problem.

1.5 The same FID was processed again, but now a strange, almost inverted copy of the spectrum is superimposed on the normal spectrum. What did the operator do wrong? This is a case of doing a common sequence of commands in the wrong order, but a very similar result is obtained if the spectrometer electronics are not set up properly.

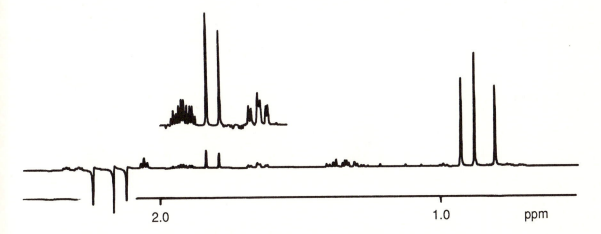

Answer A phase correction using a previous set of phase parameters was applied before the FT was performed rather than after. The Fourier transformation assumes that the two channels of data are collected in true quadrature, that is at the same gain but 90° out of phase with each other. Since phasing serves to change the phase angle between the the real and imaginary parts of the spectrum, carrying out the operation before Fourier transformation modifies the phase of the time domain data and creates quadrature images (§1.2.6 and §1.3.2). The spectrum now contains components at the correct frequency but with the wrong sign. The plot is centred on the reference frequency and we can see that each methyl singlet and its inverted image are equidistant from the centre.

Similar distortions result from an instrumental problem that is common in older spectrometers. If the data are not exactly 90° out of phase, quadrature images will appear as above. Usually, we collect data with phase cycling to suppress the quadrature images, but if the phase error is severe, signal intensity will be lost and there will still be residual quadrature images. Collection of quadrature data without phase cycling will reveal just how serious the problem really is.

1.6 The spectrum below was generated from the same FID by yet another error in data manipulation; the same effect can be caused by an error in setting up the collection of data. What are the two errors?

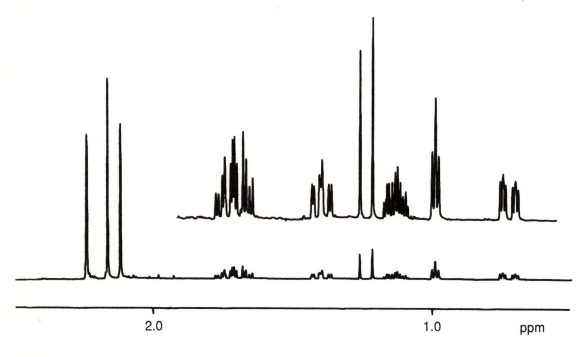

Answer The spectrum was produced by the misapplication of a trick commonly used in solid-state NMR spectroscopy (§8.3.1 and §9.7). If the early data points in the FID are distorted by the recovery time of the probe and/or receiver, one can often improve the appearance of the baseline by left-shifting the data file and adding a zero to the righthand end for each left-shift. If quadrature detection is in use, it is essential that an *even* number of left-shifts be performed. In this example, a single left-shift was applied and the frequency sense of the spectrum has been reversed; the signals have been flipped from left to right in the frequency domain because we have, in effect, reversed the x and y axes in the experiment (§1.3.2).

A large missetting of the spectrum reference frequency before collection of data leads to the same effect. If the error is large enough, the entire spectrum will be aliased across the spectral window and appear reversed. This is a particularly common occurrence when spectra are obtained from nuclei with very large chemical shift ranges. This spectrum reversal can obviously be misleading; it can also lead to a severe loss of signal intensity because filters usually suppress signals from outside the spectral window (§1.3.1).

1.7   A new FID has been acquired. On Fourier transformation, the spectrum displayed below was produced. What is the source of the extra and distorted peaks that have appeared?

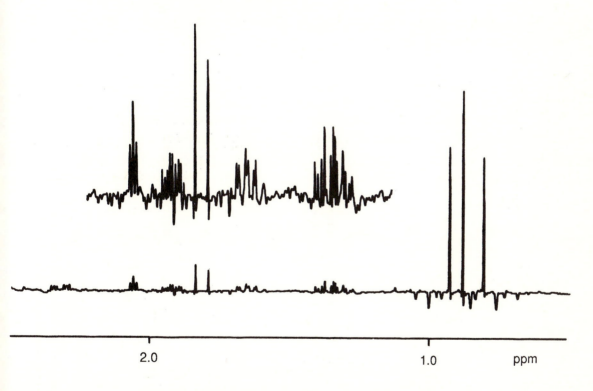

Answer The FID has been clipped by having the receiver gain too high. This problem is easily recognized by examination of the FID (§1.4.3). The early part of the FID is flat both on the top and the bottom but then begins to decay normally. If the data also contain a large dc offset signal, the FID may be flat on either the top or the bottom rather than both. The effect of clipping may also be to produce a rolling or oscillating baseline.

1.8 In an attempt to save experiment time, the operator acquired only 4K data points, zero-filled, and applied the Fourier transformation without apodization. The resulting spectrum of the methyl singlets is shown to the right. What has gone wrong?

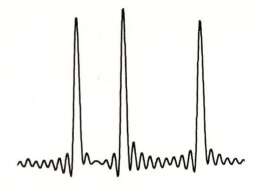

Answer The FID was truncated by the short acquisition time (§1.3.3 and §1.3.4). The sharp cut-off at the end of the FID has led to 'sinc wiggles' around each peak. This problem can be avoided either by using a longer acquisition time or by applying an apodization function that forces the FID to zero.

1.9 In an attempt to improve the digital resolution, the spectral width was reduced. The full spectrum is shown below. The methyl singlets cannot be phased with the rest of the spectrum. What has happened?

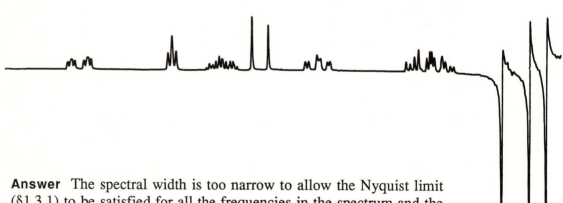

Answer The spectral width is too narrow to allow the Nyquist limit (§1.3.1) to be satisfied for all the frequencies in the spectrum and the methyl signals are 'folded' into the window. On spectrometers that use a different version of the Fourier transformation, the aliased data may appear at the other end of the frequency spectrum but will still be out of phase with the rest of the signals. Clearly, the spectral width needs to be increased.

1.10 In these spectra the spectral width was increased to ensure that there was no folding. What caused the loss of resolution?

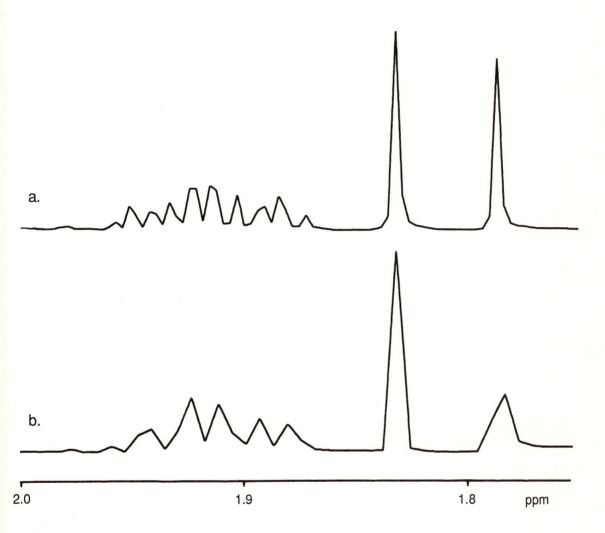

Answer The digital resolution in **a** is 1.2 Hz/point and in **b** it is 2.4 Hz/point. The loss of resolution is a result of digitally-limited data collection and has nothing to do with the field homogeneity (§1.3.1). This error is often made by users who try to set up 'standard' operating conditions to permit the running of samples in a variety of solvents. If this is attempted, the spectral width needs to be set very wide to accommodate the entire spectrum wherever the solvent deuterium lock signal comes. This way of operating throws away useful spectroscopic information; at the very least, separate standard conditions should be set for each solvent used.

1.11 While preparing this chapter, we found the 'spectrum' below on the console display. What must have happened?

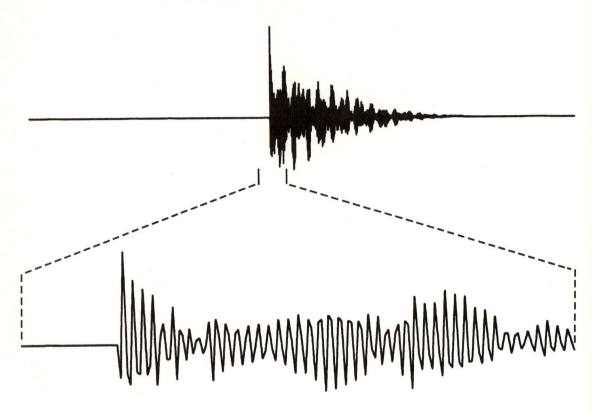

Answer Somebody typed 'FT' twice and so applied the Fourier transformation to a frequency domain spectrum. The result is a time domain spectrum with the phases of the components mixed up. It looks like an FID.

1.12 Before we acquired the spectrum to the right, the field homogeneity was carefully adjusted by maximizing the deuterium lock signal but the methyl singlets are split into distorted doublets. In fact, every line in the spectrum is identically split. What is producing this effect?

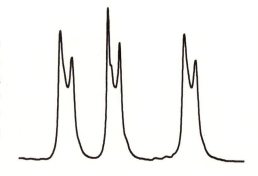

Answer This is an example of a 'split field': the apparent maximum on the lock signal can represent a local maximum in the field homogeneity. In this example, there are two regions of uniform but different field within the sample. It is possible to produce more than two regions within the sample and so obtain multiple lines for each signal. Different magnets will behave differently in this regard but for this particular magnet the problem was solved by adjusting of one of the gradients along the field axis while watching successive FIDs. The split field appears as an amplitude modulation of the FID and it is relatively easy to find a region where the modulation will vanish. Fine shimming on an FID is often the best way of obtaining the optimum field homogeneity.

1.13 To the right is part of the spectrum from a single transient obtained without the use of the deuterium lock. All the signals in the spectrum show the same pattern. What is the source of the distortion?

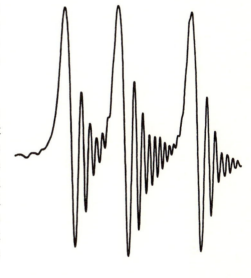

Answer The spectrometer was sweeping the magnetic field to find a lock signal, so the resonant frequencies were changing rapidly during the collection of the FID. One observes the same 'ringing' pattern when continuous wave spectra are swept too fast. If one tries to collect several transients under these conditions, one obtains many copies of the spectrum superimposed at different frequencies. The field sweep can be turned off if a lock is not going to be used.

1.14 This is part of the spectrum from 32 transients obtained without use of a lock but after the field sweep was switched off. It was collected during a busy period in the laboratory immediately after a spectrum in which the lines were very sharp. All of the lines in this example spectrum have the same shape. What is wrong?

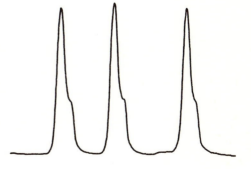

Answer Because the spectrometer was not locked, the field drifted. In this situation the movement of people and equipment in the vicinity of the magnet cause the field to vary slightly between and during the acquisitions of the spectrum. Drifting fields can cause a variety of lineshapes depending on the source of the drift. If one is simply observing long-term resistive losses in a superconducting magnet, the signals are usually flat-topped. This is because the field is very gradually drifting in a constant direction during acquisition. It is more common to observe a uniform broadening as the field wanders about some statistical average.

1.15 The ^{13}C spectra in the next few problems were collected on the same sample as the ^{1}H spectra. Each is the result of 250 transients collected with a spectral width of 25000 Hz into a 32K data file. A tip angle of $\pi/4$ was used with a relaxation delay of 6 s. The time domain data were treated with an exponential multiplication using a line broadening of 5 Hz. In all the spectra the region from 70 to 1 ppm is plotted. The solvent resonance at 77 ppm and the camphor carbonyl resonance at 220 ppm are not shown. For reference purposes, the upper spectrum opposite was collected with low power (30 dB below 20 watts) broadband ^{1}H-irradiation during the relaxation delay to build up a good NOE, and high power (5 dB below 20 watts) irradiation during the acquisition to ensure good decoupling.

The assignments for each signal are not important for the purposes of this chapter, but they are described in §1.4.9, §3.3.2, §4.2.3 and §8.5.1. The signal at 30 ppm is C_6 while that at 27 ppm is C_5. The carbon assignments are made from a $^{1}H-^{13}C$ shift correlation to the attached protons, whose assignments are proved by the NOE experiment shown in §6.5.1. We have confirmed the assignments by an INADEQUATE spectrum.

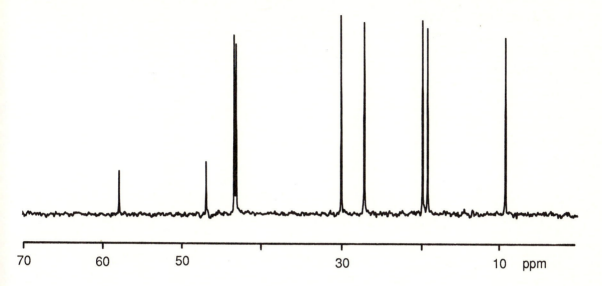

1.16 The spectrum below was collected in the same way apart from one experimental parameter. What has caused the deterioration in signal-to-noise ratio and loss of some signals?

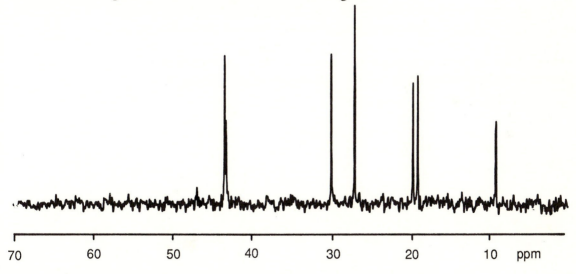

Answer The relaxation delay was set to zero so that the slowly relaxing quaternary carbon signals, which were already smaller than the other signals in the reference spectrum, were not able to recover between pulses. The whole spin system became partially saturated and signal intensity was lost. The signals that were least affected are those due to the carbons that relax most rapidly (§1.4.7, §1.4.9 and §6.4.1).

1.17 In this case, the spectrum was again obtained with only one parameter changed from the reference spectrum. What is the reason for the difference between the spectra?

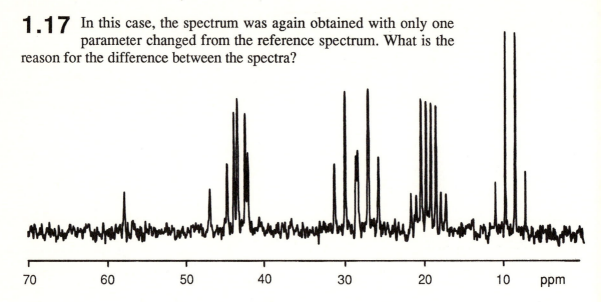

Answer The decoupler was switched off during acquisition so we have obtained the ^{1}H-coupled spectrum.

1.18 Now a different parameter has been changed, causing the peaks to become split. This is clearly seen on the expanded methyl carbon at 10 ppm; the other methyl signals look similar, while those of the methylene carbons are narrow triplets, and those of the methines are narrow doublets. What is causing the small splittings?

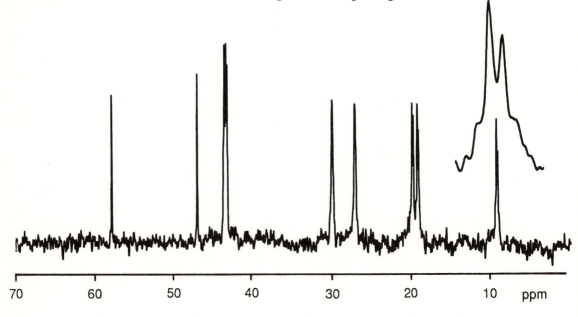

Answer The decoupler power was left at the low power setting during the acquisition. The power was not high enough to decouple the protons fully so we see the quartet splitting scaled to a smaller value by the low-power decoupling.

This used to be a common way of distinguishing methyl, methylene and methine carbons. However, as can be seen in the spectrum above, it is not a clean experiment; methylene carbons with non-equivalent protons attached often give particularly messy results (§8.2.1). *J*-modulated spin-echoes, INEPT or DEPT provide much more reliable ways of determining multiplicities (§3.3.2 and §8.5).

1.19 The spectrum below was obtained with a more drastic alteration in conditions. What can cause the non-uniform scaling of the *J*-splittings?

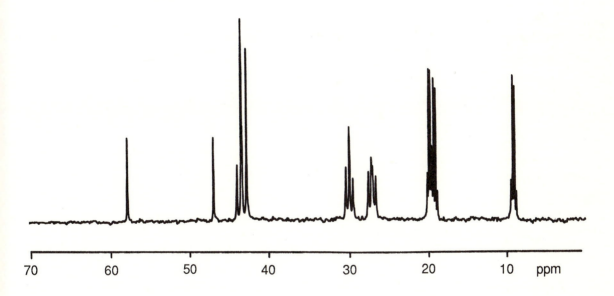

Answer The spectrum was obtained with continuous wave decoupling. The ^{1}H-decoupler was set to the low-frequency, high-field, end of the spectrum, thereby decoupling the high-field methyl protons more efficiently than the lower field methylene and methine protons.

This is the classic 'off-resonance decoupling' experiment. If the residual splitting is measured as the proton decoupling frequency is varied, then it is possible to find the chemical shifts of the protons attached to each carbon. This approach has been replaced by two-dimensional C–H correlation.

1.20 Trace **a** shows a *J*-modulated spin-echo spectrum obtained with an interpulse delay of 7.0 ms and broadband ^{1}H-decoupling after the π pulse. For trace **b** one parameter was changed. Why has most of the spectrum vanished?

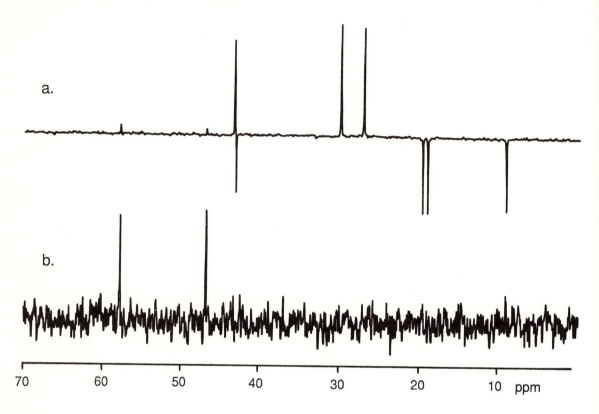

Answer The interpulse delay was set to 3.5 ms, which is $(2J)^{-1}$ for typical $^1J_{CH}$. Under these conditions the doublets, triplets and quartets all cancel, leaving only the singlets from quaternary carbons (§3.3.2).

1.21 While setting up for the previous spectra, we obtained the spectrum opposite. The spectrum has been phased to correct the phase of the highest field methyl signal. Individual signals are expanded across the spectrum so that their phases can be seen more clearly; the phase errors cannot be removed. Such phase problems may be generated in several ways, but in this case one parameter was misset in the *J*-modulated spin-echo sequence. What is the most likely source of this problem?

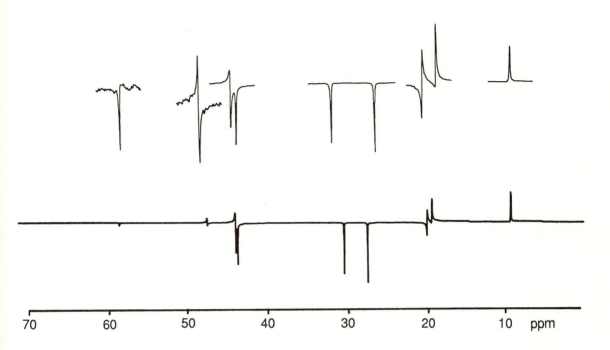

Answer The phase error in the signals is frequency dependent across the spectrum. The error was actually generated by setting to zero the width of what should have been the π-pulse. There is now no refocussing of the chemical shifts before acquisition. In the same pulse sequence, one can generate the same error by having unequal delays before and after a good π-pulse.

In general, any sequence which introduces a delay between the pulse(s) and the beginning of data collection will produce this type of phase error. Inevitably, there is always a short delay between the end of the pulse and the beginning of acquisition, but if the delay is only a few microseconds then the frequency-dependent phase error is less than 360° and is easily removed. However, in the spectrum above, the delay is 14 ms so that frequency differences of several kHz will introduce phase errors of thousands of degrees; these are virtually impossible to remove with currently available phasing algorithms.

The frequency-dependent phase errors seen above are the reason why virtually all modern pulse sequences contain refocussing π-pulses in the middle of lengthy evolution periods.

1.22   Below is the symmetrized, magnitude, COSY-60 spectrum of camphor with the one-dimensional spectrum plotted to the same scale. This spectrum was collected using the full phase cycling normally applied to suppress artefacts. Each FID consisted of 1K data points, and 256 increments were used. The data were multiplied by a sine bell window function in both dimensions to improve the appearance of the contour plot.

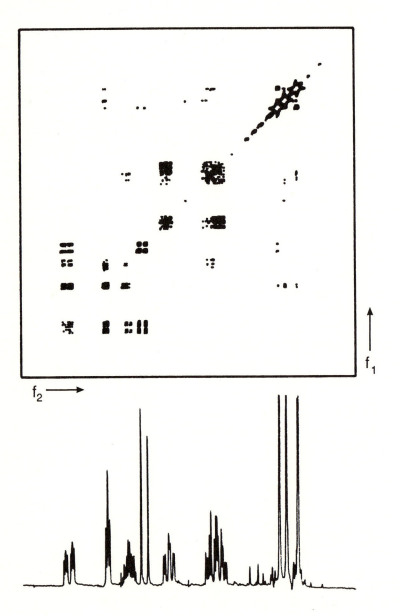

1.23 Spectrum **a** below was collected with no phase cycling at all. There are three problems with the resulting spectrum: there are strong signals running vertically, parallel to the f_1-axis; there is a horizontal ridge of signals along the line $f_1 = 0$; and there are two 'copies' of the COSY spectrum, one arranged around each diagonal. What is the source of each of these 'errors'?

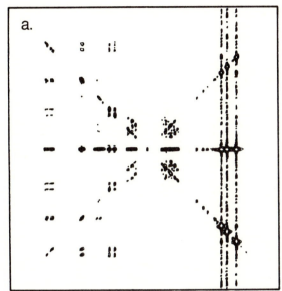 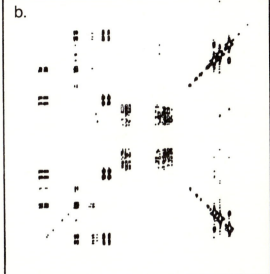

Answer The signals parallel to the f_1-axis are due to t_1-noise (§4.3.3). They arise from several sources including instability in the timing and phase of the pulses used in the sequence. The horizontal ridge is produced from the xy-magnetization that results when residual z-magnetization is tipped by the second pulse in the sequence. The second copy of the spectrum results because the pulse sequence does not distinguish between positive and negative frequencies.

Spectrum **b** above demonstrates how these problems can be separated by phase cycling. In this case, the only phase cycling applied is a simple phase alternation of successive pairs of pulses. The two sets of COSY responses remain intermingled in the spectrum but the $f_1= 0$ signals are removed and the effects of t_1-noise are substantially reduced.

The two copies of the COSY spectrum and the $f_1 = 0$ responses can all be separated without phase cycling if one is prepared to sacrifice digital resolution by increasing the f_1-spectral width. The following spectrum was collected without phase cycling, with quadrature detection OFF, and with the frequency reference offset to the low-field end of the spectrum. We can now separate in one contour plot the positive frequency or P-type COSY, the negative frequency or N-type COSY, and the $f_1=0$ components of the spectrum. This ability to separate the P and N components without phase cycling is used in some types of phase-sensitive COSY experiments.

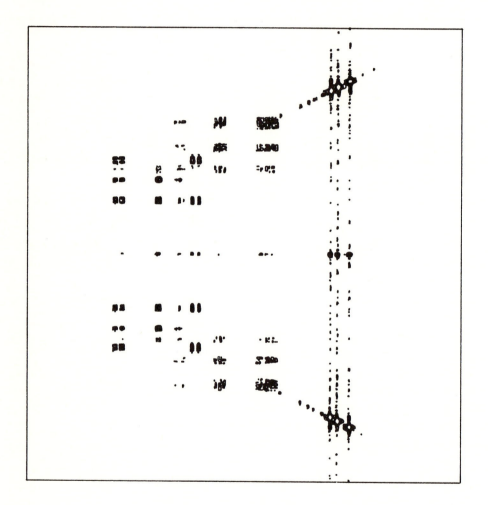

1.24 The top spectrum opposite results from using a COSY sequence, but very similar spectra can be produced using almost any two-dimensional sequence. With careful phase cycling, it is possible to to produce a remarkably uniform plot containing only two-dimensional noise. What is wrong in this sequence?

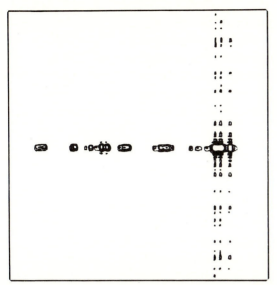

Answer The inter-pulse increment was set to zero, producing a small, constant delay rather than an incremented delay. There is no frequency evolution between successive spectra, so this is a two-dimensional plot of a one-dimensional spectrum. It is complete with t_1- noise!

1.25 Even when the operator got the phase cycling right, there was room for trouble. The spectrum below was acquired in the same way as 1.22 except that 64 increments were acquired, the data set was zero filled and no apodization was applied. Why is the t_1-noise so regular? The spectrum on the right shows the methyl region expanded.

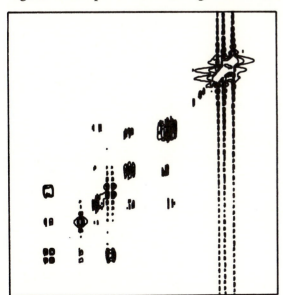

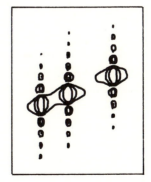

Answer The evolution in t_1 was truncated by the short t_{max} (§4.7.6), giving sinc-wiggles analogous to those in problem 1.8. In COSY spectra, this is a problem only if coarse digital resolution is used in f_1. In heteronuclear two-dimensional spectra it is almost always a problem. In both cases it is solved by appropriate apodization in f_1 (such as sine bell) or by collecting data for a longer time to give the finest possible digitization in f_1.

Conclusion

In this chapter we have outlined several examples of simple errors which can produce problems ranging from poor quality one-dimensional spectra to useless two-dimensional spectra. The errors fall into two groups. Those which result from incorrect data processing can usually be rescued by simply recognizing the problem and re-processing the original data with one or more parameters changed. Those problems which result from incorrect data-acquisition conditions are more serious because the data are 'damaged'. Even in these cases, it may be possible to salvage an experiment by applying suitable changes in the data processing; as we saw above, truncated data can often be used if a severe exponential multiplication is applied.

There are many other simple errors not covered in this chapter. For example, when the relaxation delay between acquisitions in a COSY experiment is too short, an extra diagonal with a slope of two usually appears. We have not covered a whole family of errors that produce noise and no signal. Among the most common are the many ways of arranging phase cycles such that the receiver phases do not follow the correct components generated by the transmitter phases. For example, applying quadrature phase cycling to the transmitter and not to the receiver will cancel the signals but leave the noise.

We have illustrated a set of old, and mostly well-known, errors. We have genuinely committed most of them by mistake, but, having worked through them, you should be able to avoid them. The objective when running a modern NMR spectrometer should be 'make **NEW** mistakes'.

2 Interpretation of spectra

E2.1 When this cyclohexanone is dissolved in benzene, $J_{AB} = 3$ Hz, but when it is dissolved in methanol, $J_{AB} = 11$ Hz. What are the conformations in these two solvents? Why are they different?

Answer The large J_{AB} of the methanol solution is only compatible with both protons being axial. If we assume that the ring is a chair, then the conformation must be as shown in **a**; this is as expected, with the large *iso*-propyl group being equatorial. The small J_{AB} in benzene solution indicates that both protons are now equatorial; the ring has flipped into the other chair form, **b**.

a   b

The polar hydroxyl group needs to be stabilized by hydrogen bonding. This is readily achieved by bonding to solvent molecules in methanol, but in benzene there can be no stabilizing interaction with the solvent. However, intramolecular hydrogen-bonding to the carbonyl group is possible after the molecule has flipped into the other chair form, **b**, bringing the hydroxyl group into a suitable orientation. The energetic cost of making the *iso*-propyl group axial is reduced by the absence of 1,3-diaxial interactions at the carbonyl group, and is fully paid for by the benefit of hydrogen-bonding the hydroxyl group.

E2.2

The ^{199}Hg spectrum of a solution of $[Hg_3][AsF_6]_2$ in liquid SO_2 (200K, 44.8 MHz) is shown below. Assuming the $[Hg_3]^{2+}$ cation is linear, how may the spectrum be interpreted?

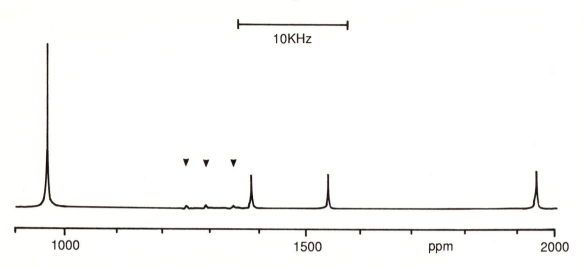

Answer ^{199}Hg is 16.84% abundant and has $I = {}^1/_2$; it is represented as *Hg below. The major NMR-active isotopomers present will be $[^*Hg-Hg-Hg]^{2+}$ and $[Hg-^*Hg-Hg]^{2+}$, and we would expect to see resonances due to the terminal and central mercury atoms in a ratio of 2:1. These are the resonances at -965 and -1968 ppm respectively.

However, we must also consider the minor isotopomers containing two *Hg atoms, $[^*Hg-^*Hg-Hg]^{2+}$ and $[^*Hg-Hg-^*Hg]^{2+}$. The latter is not detected because it contains two equivalent nuclei whose resonances will be coincident with those of $[^*Hg-Hg-Hg]^{2+}$. In contrast, the non-equivalent spins in the remaining isotopomer, $[^*Hg-^*Hg-Hg]^{2+}$, are expected to couple strongly with each other. The direct $^*Hg-^*Hg$ coupling (139 600 ± 1000 Hz!) is large compared to the difference in chemical shifts (~45 000 Hz), so an AB spin system is formed. The lines at ~-1400 and ~-1550 δ are the intense inner transitions of the AB spin system, while the two outer transitions are well outside the spectral width and are not observed. This is believed to be the first observation of direct $^*Hg-^*Hg$ coupling.

Close study of the spectrum reveals low intensity resonances, marked with arrows. These may be assigned to the low-abundance AB_2 spin system $[^*Hg-^*Hg-^*Hg]^{2+}$.

[Gillespie, R.J., Granger, P., Morgan, K.R., and Schrobilgen, G.J. (1984). *Inorg. Chem.*, **23**, 887–91.]

2.1 The 400 MHz ^{1}H spectrum of a sample labelled 'CD$_2$Cl$_2$' is shown here. Assign and explain the signals, which have a chemical shift of around 5.2 ppm. The scale markers are 5 Hz each.

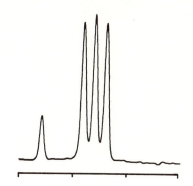

2.2 Explain the appearance of this titanium spectrum of a solution of TiCl$_4$, obtained at 22.55 MHz in a 9.4 T magnet. The scale markers are 1000 Hz each.

2.3 Give complete assignments for the non-aromatic carbons of this compound. The multiplicities refer only to C–H splittings.

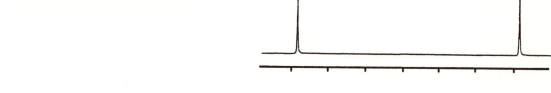

21.9	q	
41.3	t	
51.8	d	
61.0	d	$J_{CP} = 161$ Hz
65.8	d	Attached to a ^{1}H signal at δ 4.19 (qn, 6.5 Hz)
68.5	d	Attached to a ^{1}H signal at δ 4.01 (dt, 2 and 7 Hz)
69.0	t	$J_{CP} = 7$ Hz
69.3	t	$J_{CP} = 7$ Hz
173.2	s	
207.3	s	

2.4 Assign and interpret the major features of the two multiplets from the ¹H-decoupled ¹⁹⁵Pt spectrum of this complex. The bar represents 1000 Hz for both multiplets.

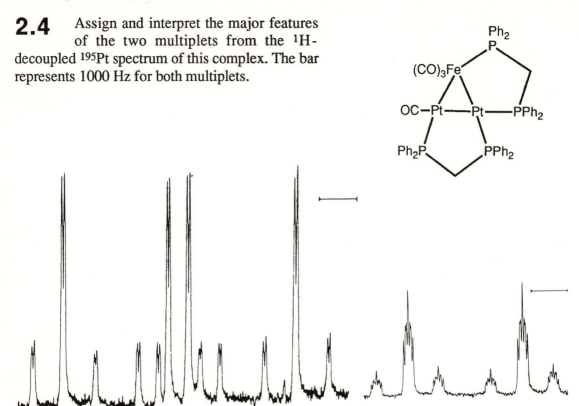

E2.3 Assign the 220 MHz ¹H spectrum of B_4H_{10} as fully as possible. The protons coupled to ¹⁰B are broad and not observed.

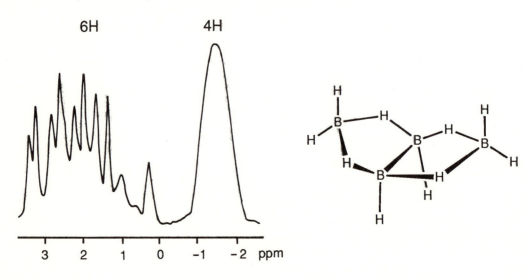

Answer There are four distinct proton environments. These are axial or equatorial terminal sites on the wingtip BH_2 groups, terminal sites on the hinge BH group and bridging sites.

As usual, ^1H–^1H coupling is lost in the broad lines of boranes. The bridging hydrides are seen as the broad unresolved multiplet centred at -1.38 ppm. The width of this signal may be associated with B–B coupling (a second-order effect) or with some fluxional process. It is typical of bridging B–H–B resonances. The three remaining terminal hydride signals should each appear as a 1:1:1:1 quartet as a result of coupling to ^{11}B (we may ignore the 20% ^{10}B present). This accounts for the observed pattern, in which 11 of the expected 12 signals are resolved. Analysis of the intensities and coupling constants in the spectrum allows the separate sub-spectra to be assigned as shown below, although the axial and equatorial sites of the wing-tip position may not be distinguished. ^{11}B decoupling experiments support these assignments.

Position	Shift	J_{BH}
Bridge	-1.38	–
Wingtip 1	2.26	134 Hz
Wingtip 2	2.46	125 Hz
Hinge	1.34	155 Hz

[Leach, J.B., Onak, T., Spielman, J., Rietz, R.R., Schaeffer, R., and Sneddon, L.G. (1970). *Inorg. chem.*, **9**, 2171–5.]

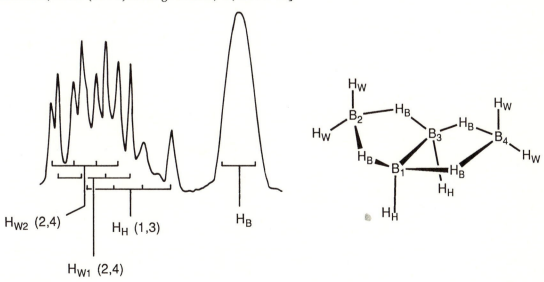

2.5 Assign and interpret the 300 MHz ^{1}H NMR spectrum of this compound as fully as possible. Can you draw any conclusions about the conformation ? [The expanded multiplets A–G are all plotted to the same height].

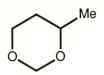

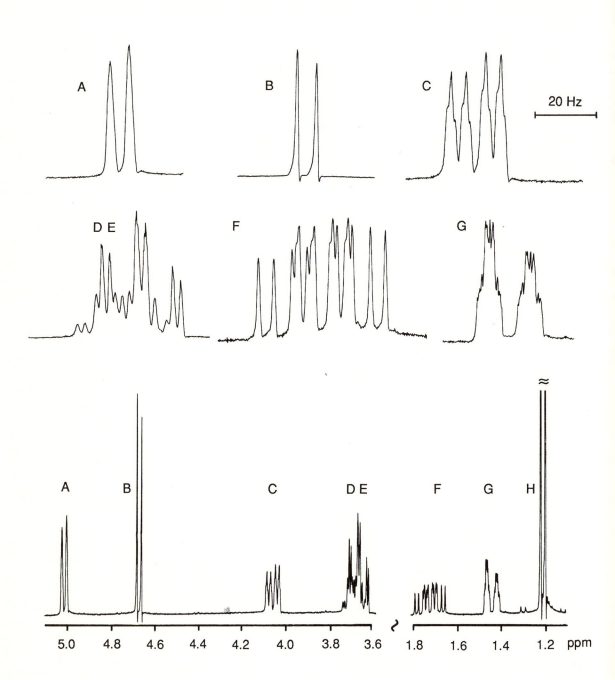

E2.4 The observed ^{1}H–^{1}H NOEs in β-pinene are extremely small, even after very long irradiation times, while the resonances in its natural abundance deuterium spectrum are very sharp. What single property of the molecule accounts for both of these effects?

Answer The β-pinene molecule is small, roughly spherical and lacking in polar functional groups. It can therefore tumble very rapidly in solution without perturbing surrounding solvent molecules. This rapid tumbling leads both to inefficient proton relaxation by the dipole–dipole mechanism, so that NOEs are too small to measure easily (§6.2.1 and §6.4.3), and to relatively slow deuterium relaxation, which gives sharp deuterium lines (§7.2.3).

2.6 The ^{1}H spectrum of this ferrocene contains the three signals listed below. Assign these signals using only the relaxation rates given and comment on the relative rotation rates of the two rings. You can assume that the side-chain has no components that induce relaxation.

A	2H	m	$R_1 = 0.26$ s^{-1}
B	2H	m	$R_1 = 0.42$ s^{-1}
C	5H	s	$R_1 = 0.19$ s^{-1}

E2.5 Assign as far as possible the COSY spectrum of the aliphatic region of this tripeptide analogue (over the page). Conditions: 400 MHz, D$_2$O, magnitude spectrum with coarse digital resolution.

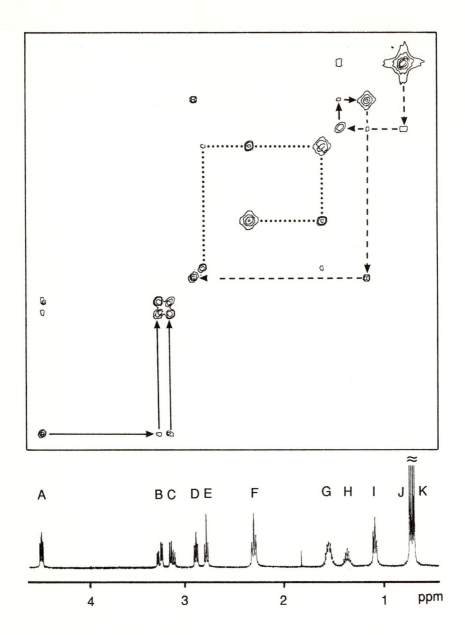

A   B C D E   F   G H I J K

4   3   2   1   ppm

Answer As the spectrum was acquired in D_2O, the NH protons will have exchanged. The aliphatic region should therefore only contain responses from three distinct spin systems: CH_2CH_2CH from the lefthand portion, $(CH_3)_2CHCH_2CH$ from the centre of the molecule and CH_2CH from the righthand side. Each of these should give a separate set of cross-peaks. Inspecting the spectrum, it is fairly easy to find these three sets of signals: they are indicated by solid, broken and dotted lines. But which is which? We need a starting point for each spin system.

One obvious start point is the pair of intense methyl doublets J and K which must belong to the central fragment. These have very similar shifts but are diastereotopic (non-equivalent) because the molecule is chiral. Following the broken line down from the JK-resonances we find a cross-peak that tracks horizontally across to the signal from H. From H we can follow the arrows to I, a two-proton multiplet corresponding to a CH_2 group, and then to D, a one-proton triplet.

The three resonances A, B and C are all coupled to each other: a solid line joins A to B and C but there are also cross-peaks between B and C, close to the diagonal. This isolated three-spin system belongs to the tryptophan residue at the right of the molecule. On chemical shift grounds, A must be the α-proton between the amide nitrogen and carboxylic acid groups. B and C are the non-equivalent geminal pair next to the indole ring.

The starting point for the final, left-hand portion of the molecule is the one-proton triplet, E. Following the dotted line, we find that E is coupled to G (a complex multiplet) which is coupled in turn to F (a two-proton triplet).

All the assignments are summarized on the structure below. The COSY spectrum has allowed us to assign the spectrum without measurements of the couplings and without reliance on subtle chemical arguments. In some of the more complex problems that follow, these additional sources of information will also be needed. The assignments given here are incomplete because we have not assigned which methyl group is which or which methylene proton is which. These are much more difficult to assign and differentiating them requires a knowledge of the stereochemistry of the chiral centres combined with information from NOE or biosynthetic labelling experiments.

2.7   Given the COSY spectrum and the assignments of H_1 and H_{11} of this compound, assign the remaining ^1H aromatic signals.

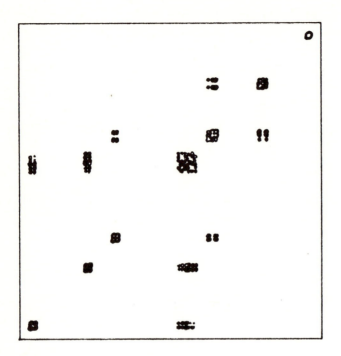

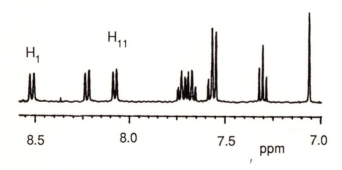

2.8   The hydride region of the ^1H spectrum of the osmium cluster $[H_3Os_3(CO)_9CBr]$ consists of a single sharp resonance flanked by low intensity 1.5 Hz doublets at ±15 Hz either side. Account for this observation.

2.9 Assign fully the ^{1}H spectrum of this aromatic compound using the COSY and NOESY plots below. The COSY evolution time was very short, so the only visible cross-peaks are due to ortho connections.

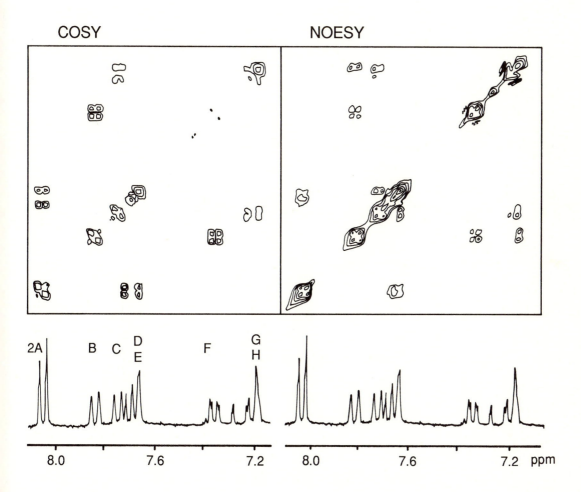

2.10 Assign as much as possible of the ^{1}H and ^{13}C spectra of catechin using the one-bond C–H correlation spectrum below. Several quaternary carbons in the 120–160 ppm region are not shown. S = solvent.

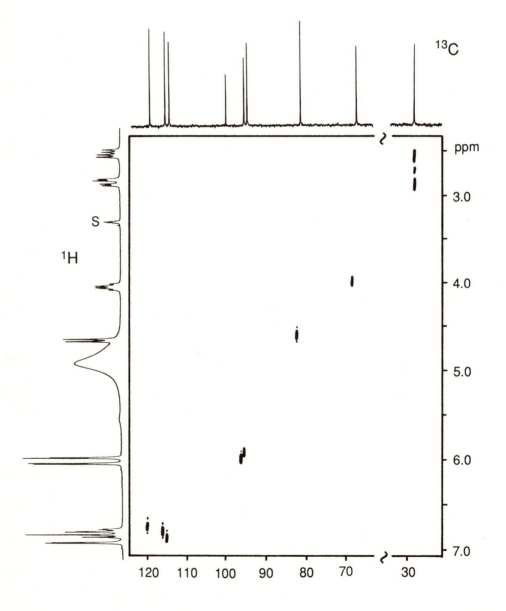

2.11 The aromatic region of the ^{13}C spectrum of this compound is shown below. The spectrum was obtained with $\pi/2$ pulses and no relaxation delay. Explain the relative heights of the signals.

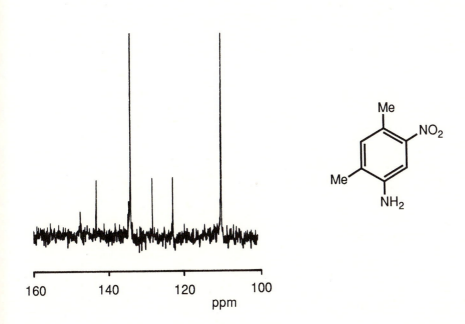

2.12 Use the COSY and one-bond C–H correlation spectra on the following two pages to assign the ^1H and ^{13}C spectra of Rogiolenyne A:

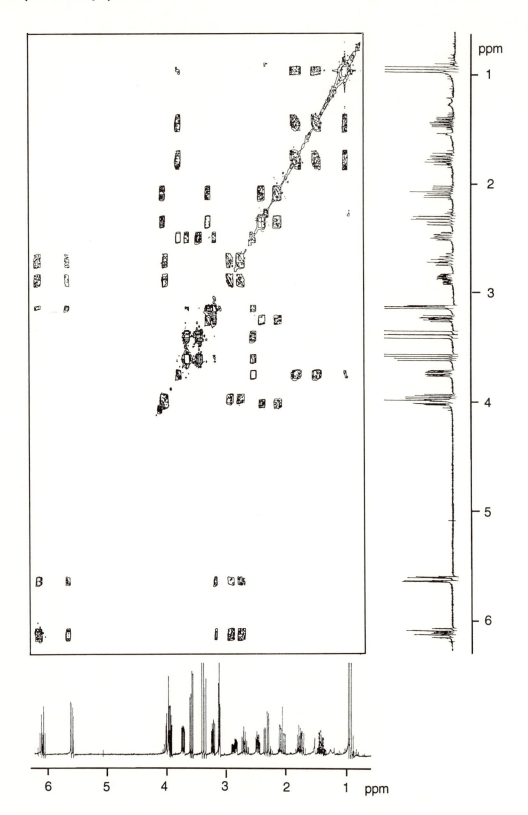

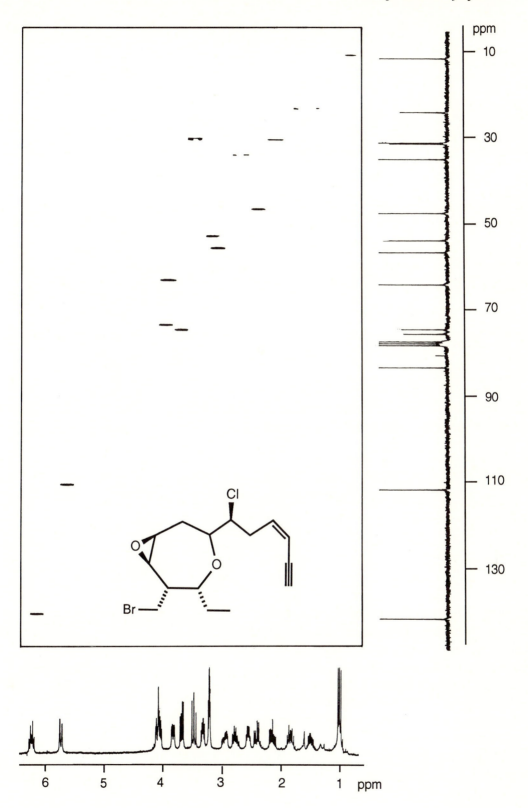

2.13 Interpret these ¹H and COSY spectra of *cis* [Ir(bipy)₂Cl₂]Cl. The signals marked X are 5 Hz doublets.

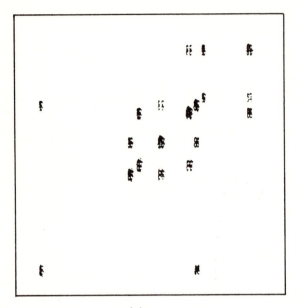

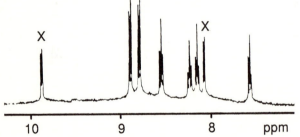

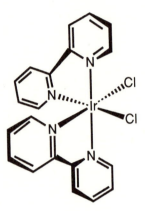

2.14 The saturated ring of this phosphate ester has four proton resonances which are shown opposite in **a**. Because the multiplets are so complex, a two-dimensional *J*-resolved spectrum was also acquired. **b** shows the f₁-slices (cross-sections) from the *J*-spectrum and **c** shows the contour plots. Assign the multiplets and explain their appearance.

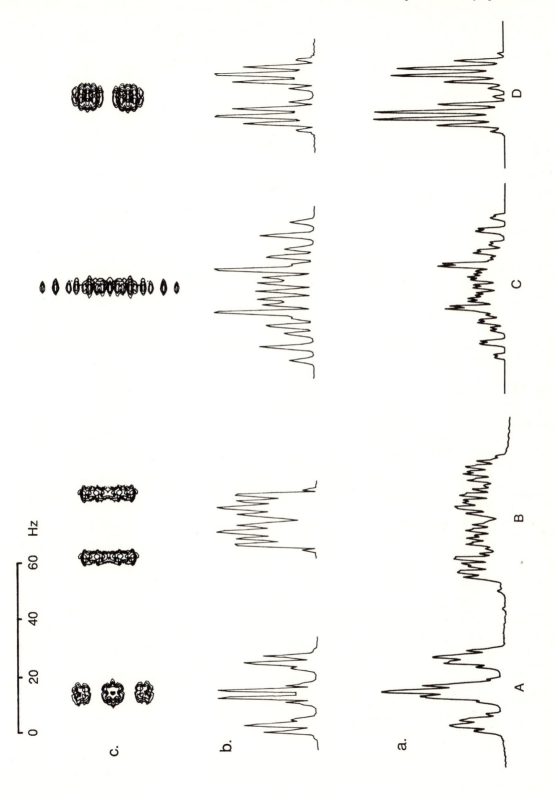

2.15 These partial 20 MHz ^{13}C spectra of cholesterol were acquired under the same conditions ($\pi/2$ pulses, short acquisition time, no relaxation delay) apart from the temperature. Explain the difference in appearance.

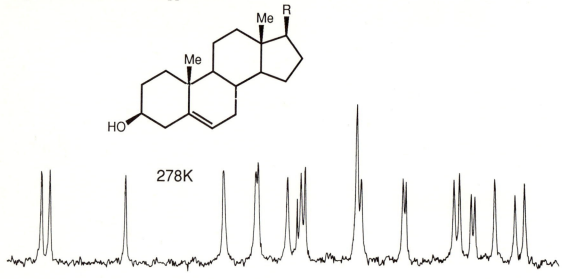

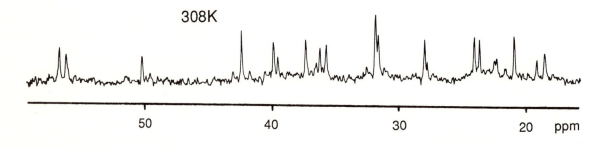

2.16 Shown opposite is part of a long-range C–H correlation spectrum of the porphyrin illustrated. Use the ^{1}H assignments to obtain a complete assignment of the quaternary carbons. The small multiple peaks around 136 ppm are due to pyridine-d_5 in the solution. Conditions: The 400 MHz spectrum was acquired with ^{1}H detection (§4.2.3), and with delays set for two- and three-bond C–H couplings. Only the quaternary carbon region is shown in f_1. Correlations to the methyl protons (within the inset box at 3.6–3.8 ppm) were much more intense and were plotted with high level contours only.

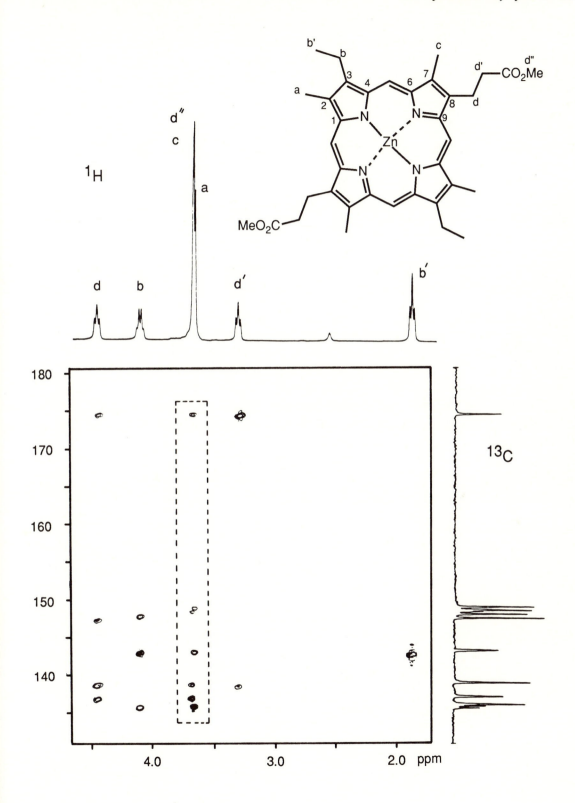

2.17 Use the table of observed NOEs for this fluoro compound to determine the dominant conformation of the amide side-chain. A zero indicates an NOE of less than 1%.

{ }	NOE observed			
	H_3	H_6	NH	Me
H_3	–	0	-23	0
H_6	0	–	0	0
NH	-5	0	–	+3
Me	0	0	+7	–

2.18 The ^{1}H one-dimensional and COSY spectra of triptolide chlorohydrin are shown opposite and selected NOEs are summarized below; strong NOEs are underlined, and weak NOEs are in parentheses. Assign the spectrum as fully as possible and determine the major conformation of ring A. Conditions: CDCl$_3$, 400 MHz, 0.8–4.8 ppm. The COSY was double-quantum-filtered and phase-sensitive; it was acquired under high resolution conditions that give correlations even for very small couplings.

{ }	Proton enhanced
C	Q
D	$\underline{C}$ N O
F	E R
G	Saturation transfer to water
M	(E I) L
O	$\underline{D}$ I J $\underline{N}$
P	K M N
Q	$\underline{C}$ $\underline{H}$
R	$\underline{F}$ $\underline{H}$
S	Saturation transfer to water

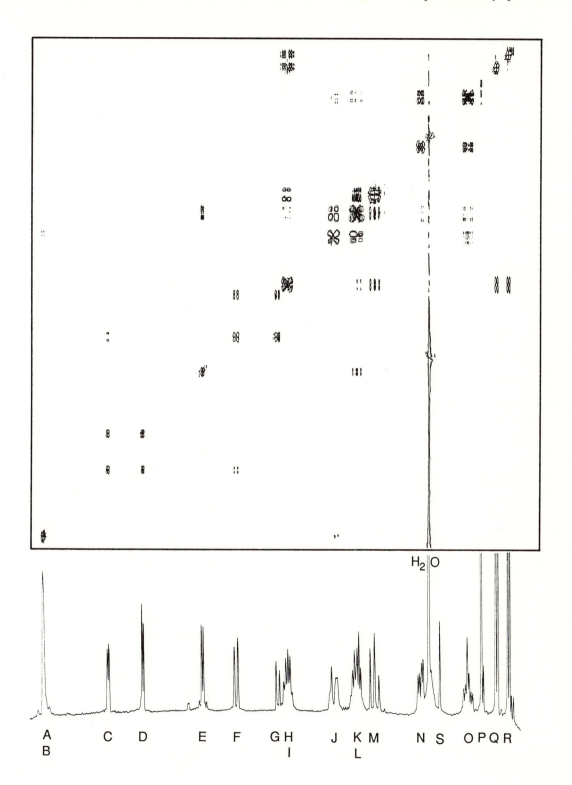

2.19 Assign as much as possible of the ^1H and ^{13}C spectra of menthol using the ^{13}C INADEQUATE spectrum below and the one-bond C–H correlation opposite.

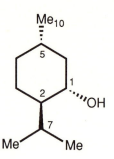

Conditions: CDCl$_3$ solution, 300 MHz ^1H and 75 MHz ^{13}C spectra. The spectrum below was obtained with a modified pulse sequence that gives a symmetrical spectrum. The cross-peaks contain doublets because the spectrum is detecting coupled ^{13}C–^{13}C pairs.

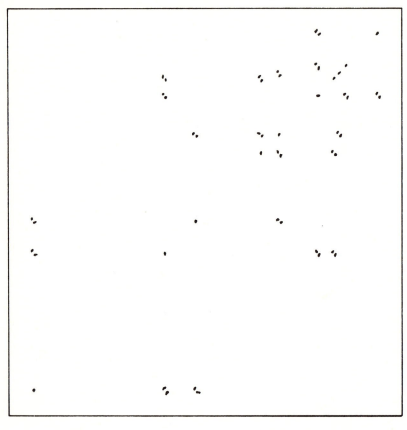

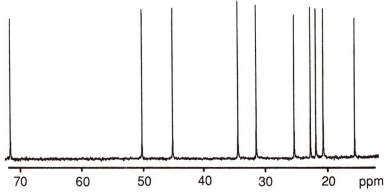

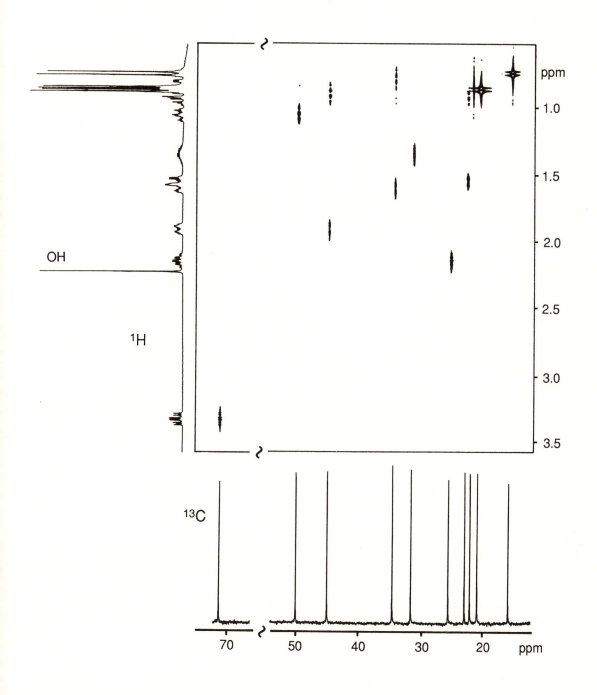

3 Symmetry and exchange

E3.1 Explain why the deuterium-decoupled ^1H spectrum of cyclohexane-d_{11} contains only a sharp singlet at room temperature but two singlets at 170K.

Answer One would expect the ring to take up the most stable chair conformation, so the lone proton will be found either in an equatorial position (**a**) or in an axial one (**b**). The two forms will be present at effectively the same concentrations because any isotope effect on the conformational preference should be too small to be significant. All the couplings to deuterium have been removed by decoupling, so the proton resonance should be a singlet.

a   b

At low temperatures, the interconversion of the two forms is slow on the chemical shift timescale (§7.2) so we see separate signals for both conformers. At high temperatures, ring flipping is interchanging the two conformations rapidly on the chemical shift timescale and we see only a single, averaged signal.

3.1 Explain the temperature dependence of the ^1H spectrum of azetidine. Both spectra were acquired with decoupling of the NH proton.

303K			243K		
3.3	4H	t	3.6	2H	m
			3.0	2H	m
2.2	2H	qn	2.4	1H	m
			1.8	1H	m

3.2   At room temperature, the carbonyl region of the 20 MHz ^{13}C spectrum of this *N*-glucoside shows a single resonance at 170 ppm for C*, a single resonance for C† at 138 ppm, and four resonances for the carbonyl groups attached to the oxygens of the glucose portion of the molecule. When the solution is cooled to 100K, the resonances for C* and C† each split into two signals of equal intensity but the remaining resonances are unchanged. Explain these observations.

$$R = C_6H_5C=O$$

E3.2   The ^{19}F spectrum of IF$_5$ at 200K contains a doublet (4F) and a quintet (1F); at higher temperatures, the spectrum collapses to a 5F singlet. What is the structure of IF$_5$ and what is happening to the spectrum on warming?

Answer   Electron counting indicates that IF$_5$ possesses one lone pair of electrons in the valence shell. We can therefore predict a structure based on an octahedron with the lone pair occupying one of the vertices. There are two types of fluorine, occupying equatorial (F$_E$) and axial (F$_A$) sites, in a ratio of 4:1. The axial fluorine is coupled to four equivalent F$_E$ atoms and appears as a quintet while the equatorial fluorine atoms appear as a doublet. On warming, the axial and equatorial fluorine sites exchange rapidly, presumably by pseudorotation, and a single resonance is observed.

3.3   Account for the appearance of the ^1H and ^{11}B resonances of a solution of this [B$_3$H$_8$]$^-$ anion. The protons coupled to ^{10}B give rise to the broad hump below the sharper signals, and can be ignored.

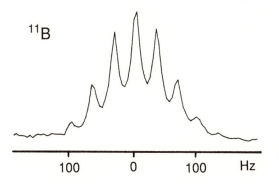

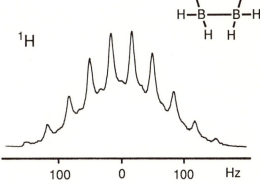

3.4   ^{15}N-Labelled ammonium chloride (NH$_4$Cl) was dissolved in a mixture of H$_2$O and D$_2$O and then strongly acidified. Interpret as fully as possible the resulting ^{1}H spectrum shown below. Conditions: 400 MHz; only the NH region is shown.

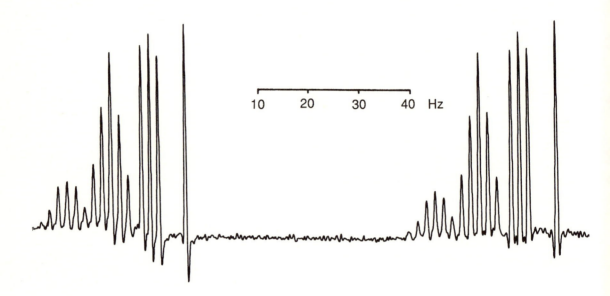

3.5   When (PhCH$_2$)$_2$NMe is dissolved in basic D$_2$O, the CH$_2$ protons of the benzyl group resonate as a singlet in the ^{1}H NMR spectrum. After acidification of the solution with DCl, these protons appear as an AB quartet. Explain these observations.

3.6   The ^{13}C spectrum of the tripeptide glutathione at neutral pH in H$_2$O solution has four signals in the 160–180 ppm region. When the spectrum is acquired in a 1:1 H$_2$O:D$_2$O mixture, two of the signals become apparent doublets. Explain why.

3.7   At low temperatures and low concentrations, the ^{1}H spectrum of this compound contains two methyl singlets. Raising *either* the temperature *or* the concentration leads to coalescence into a single methyl resonance. Why?

3.8   Explain and assign these 100 MHz ^{1}H spectra of this protected sugar. Spectrum **a** was obtained in dry DMSO-d_6 solution; **b** is of the same sample after addition of a trace of HCl gas. The broken lines show the positions of changes.

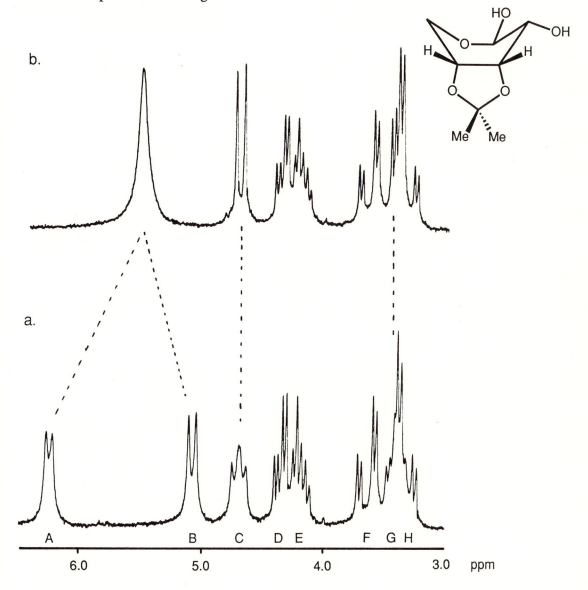

3.9   The ¹H spectra to the right are from the
methyl signals of dimethyl formamide
(HCONMe₂) dissolved in DMSO-d_6 at 393K. The
spectra were recorded on three different
spectrometers, **a** at 60 MHz, **b** at 200 MHz, and **c**
at 400 MHz.   The temperatures were very
carefully checked and are all genuinely within 1K
of the correct value; the magnets were properly
shimmed.   The spectra are plotted to the same
scale in ppm/cm, the scale markers being 0.1 ppm
apart. Why are the linewidths so different?

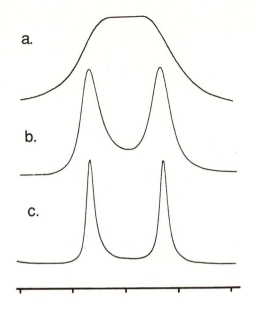

a.

b.

c.

3.10   Explain why the ¹H NMR spectrum of this biphenyl
compound contains an AB quartet centred at δ 3.4.

3.11   Why does addition of enantiomerically pure [S]-[+]-O-acetyl-
mandelic acid **a** to a benzene solution of the racemic amine **b**
lead to a doubling of the amine CH and CH₃ signals in the NMR
spectrum? Both compounds are recovered unchanged from the solution.

a

b

3.12 The solid-state CPMAS and solution ^{13}C spectra of this porphyrin are shown below. Give possible explanations for the difference in appearance. **C** is CDCl$_3$ in the solution spectrum; **P** is a crystalline polyethylene chemical shift standard added to the solid.

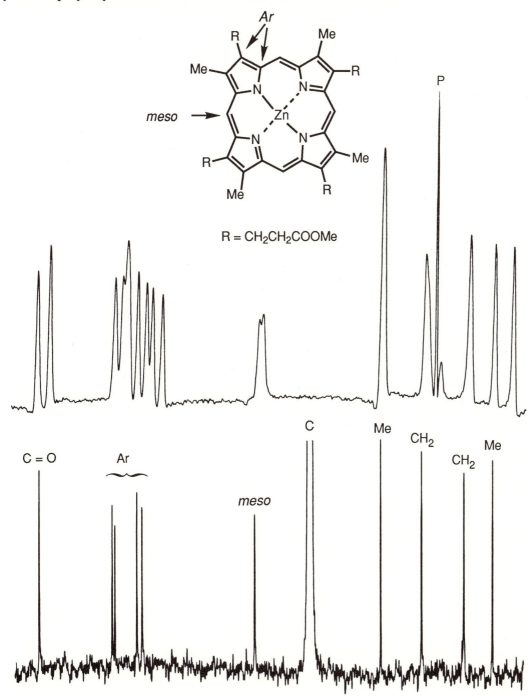

R = CH$_2$CH$_2$COOMe

E3.3 Shown below are **a** the 250 MHz ^{1}H spectrum of this pyridine derivative and **b** the NOE difference spectrum resulting from pre-irradiation of H$_{3'}$. What conclusions can you draw about the stereochemistry, conformation and bond rotations of this molecule?

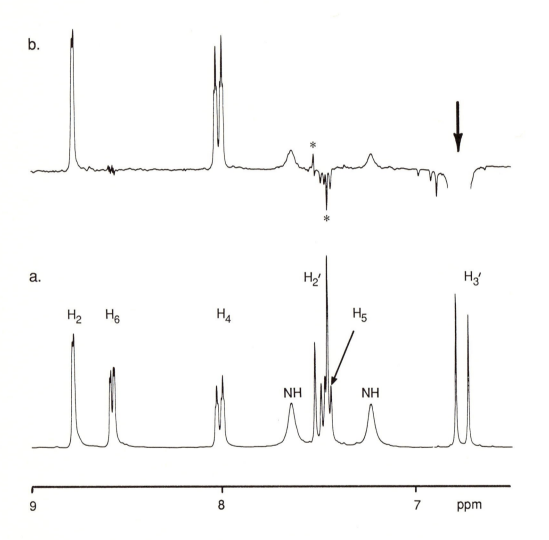

Answer $J_{2'3'}$ is 15 Hz, so the double bond is *trans*. This is probably confirmed by the lack of NOE from H$_{3'}$ to H$_{2'}$, but beware of over-interpreting the lack of NOE. The up–down H$_{2'}$ lines (*) are selective population transfer responses (§3.4 and §6.4.3) resulting from imperfect saturation of H$_{3'}$.

On average, the exocyclic double bond is likely to be coplanar with the aromatic π-system, giving two preferred conformations, **a** and **b**. These coplanar conformations correspond to potential wells separated by a barrier. The NOEs from $H_{3'}$ to H_2, H_4 and H_5 are roughly 8, 12 and -2%. The large NOE to H_4 indicates that conformation **a** is important, bringing $H_{3'}$ and H_4 close together. This is confirmed by the negative NOE at H_5 which is an indirect 'three-spin' effect resulting from the positive enhancement of H_4 (§ 6.2.3). We would expect H_5 to be equally relaxed by H_4 and H_6, so the NOE from H_4 could be as large as 25%. This makes the maximum possible negative enhancement from $H_{3'}$ 3% (12 x 25%); the observed effect is -2% with only a relatively short irradiation, telling us that the $H_{3'}$–H_4–H_5 arrangement is effectively linear in **a**.

A large enhancement of H_2 from $H_{3'}$ would not have been expected for conformation **a** because H_2 is much closer to $H_{2'}$. So there must also be a large population of conformation **b**. We cannot easily quantify the relative populations from this NOE difference spectrum because H_2 and H_4 are in such different proton environments (§6.5.3). Equilibration between the two conformations is obviously fast on the chemical shift timescale. The appropriate experiment here is separate irradiation of H_2 and H_4 and measurement of the relative enhancements at $H_{3'}$ and $H_{2'}$. This experiment is illustrated below, and the resulting spectra are shown on the next page.

As predicted, irradiation of H_2 enhances the signals of both $H_{3'}$ and $H_{2'}$. Irradiation of H_4 enhances the $H_{3'}$ and $H_{2'}$ signals; the latter is partly obscured by enhancement of H_5. $H_{3'}$ experiences a 12.5% NOE from H_4, and a 7% NOE from H_2, giving a population ratio for **a:b** of just less than 2:1.

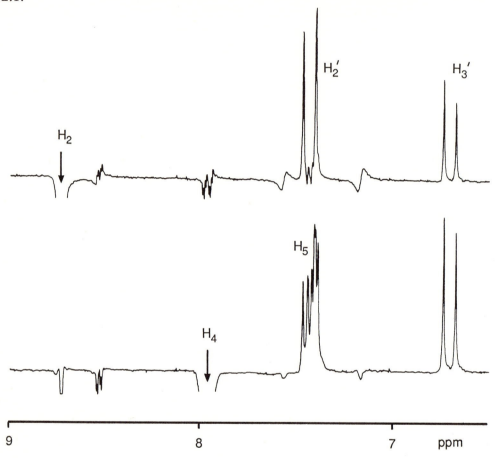

The two amide NH signals were equally enhanced in the first NOE difference spectrum (**b**, p. 64). The presence of two signals shows that they are not exchanging rapidly on the chemical shift timescale and structure **c** shows that they would be expected to be different distances from $H_{3'}$. If there is no exchange between the NH positions then irradiation of $H_{3'}$ should give an indirect negative NOE to one NH *via* the other. So we conclude that they are in rapid exchange with each other on the T_1-timescale, presumably by rotation about the C–N bond.

The alternative interpretation, that the NH protons are equidistant from $H_{3'}$ because the NH_2 group is perpendicular to the conjugated carbonyl system, seems unlikely on chemical grounds.

3.13 The reaction of osmium atoms with benzene in the vapour phase produces the yellow compound [$OsC_{12}H_{12}$]. The signals appearing in its ^1H spectrum are listed below, and the ^1H–^1H exchange spectrum of the compound is shown on the right. What exchange process is occurring?

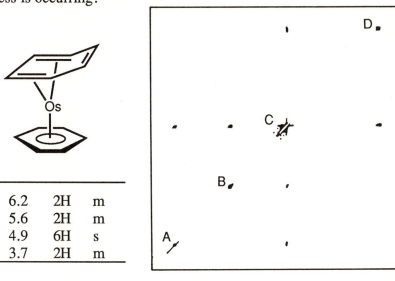

A	6.2	2H	m
B	5.6	2H	m
C	4.9	6H	s
D	3.7	2H	m

3.14 Explain why (i) the ^{13}C spectrum of **a** shows 7 resonances (excluding the X substituent) at room temperature, (ii) on warming or on addition of a *catalytic* quantity of acid the spectrum collapses to 4 resonances and (iii) the T_1-value for both C_8 and C_9 at room temperature is 21 s, even though C_8 is near an NH proton and C_9 is not.

For comparison, the ^{13}C spectrum of **b** at all temperatures shows 8 resonances excluding X, and the T_1-values for C_8 and C_9 at room temperature are 32 and 33 s respectively.

a R = H
b R = CH_3

3.15 Explain the appearance of the multiplets in the 300 MHz ^{1}H spectrum of this compound.

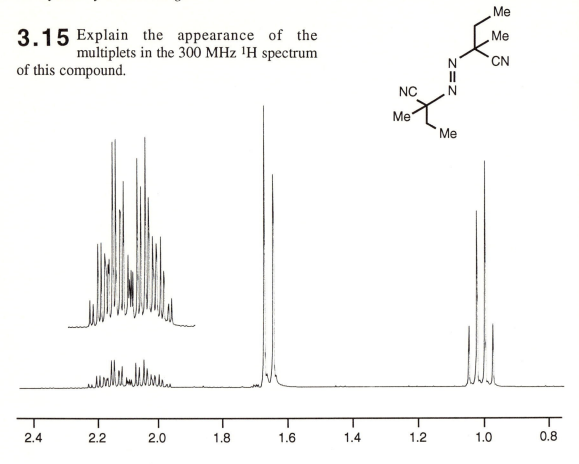

3.16 On the opposite page are shown the 300 MHz ^{1}H spectra of these compounds; **a** and **c** are dissolved in DMSO-d_6, while **b** and **d** are dissolved in CDCl$_3$. The aromatic regions are not shown. Explain the increasing complexity of the spectra.

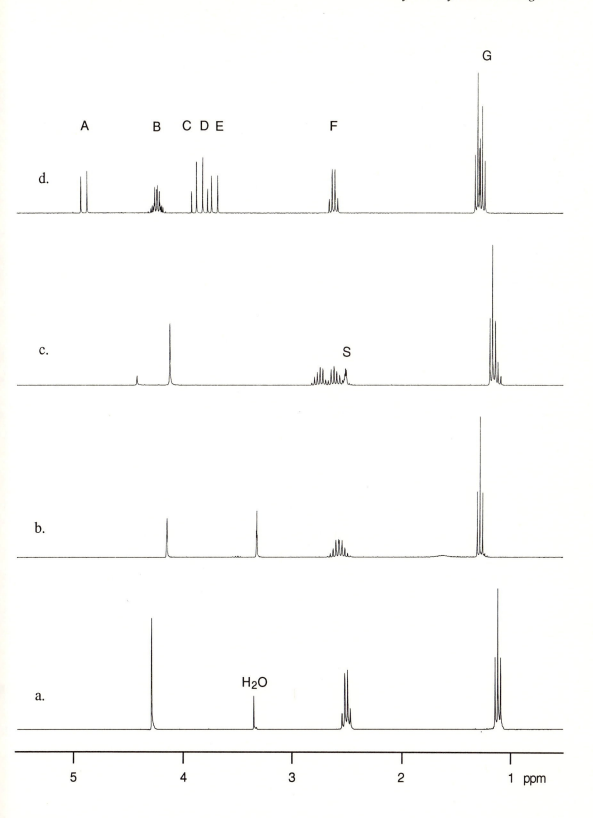

3.17 Assign and explain the aromatic part of the spectrum shown below for this compound in CDCl$_3$.

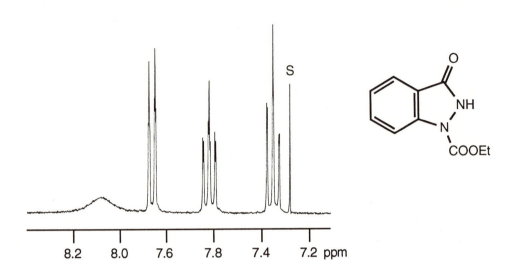

3.18 The four spectra on the opposite page were run in CDCl$_3$ which had been passed through dry alumina. There are spectra of 1,9-nonanediol, HO(CH$_2$)$_9$OH, at 13 mM and 73 mM, and of 1,3-propanediol, HO(CH$_2$)$_3$OH, at 6 mM and 110 mM.

(i)   Assign and explain the appearance of spectra of the two diols at their lower concentrations.

(ii)   Explain the differences caused by the increased concentration of nonanediol in the lower pair of spectra.

(iii) Why does signal A for propanediol at high concentration look different from that of nonanediol at high concentration?

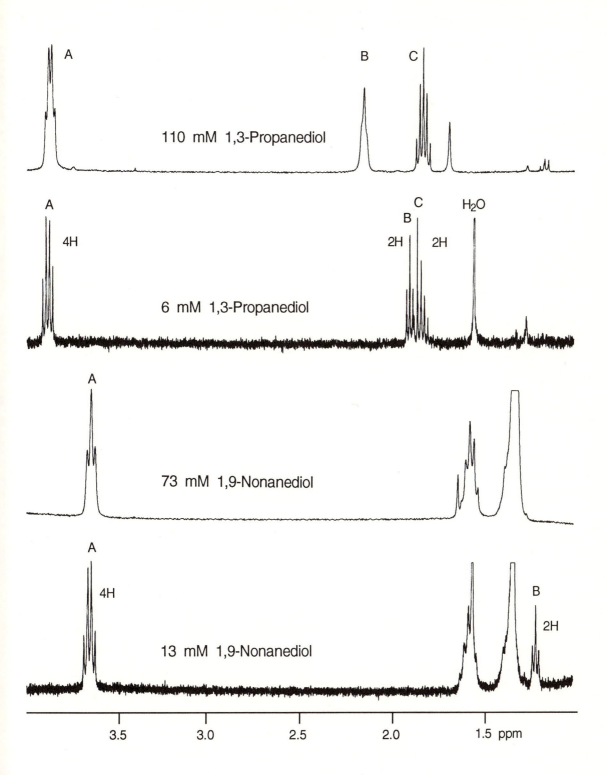

3.19 The molecule $RuH_2(CO)(PPh_3)_3$ is fluxional. The hydride protons are inequivalent and exchange with each other. There are two types of PPh_3 which also exchange with each other but not with added PPh_3. There is no intermolecular exchange of CO. Shown below are an EXSY spectrum for the hydride region and an expansion of the cross peak. Deduce the structure of the complex and the mechanism of the exchange.

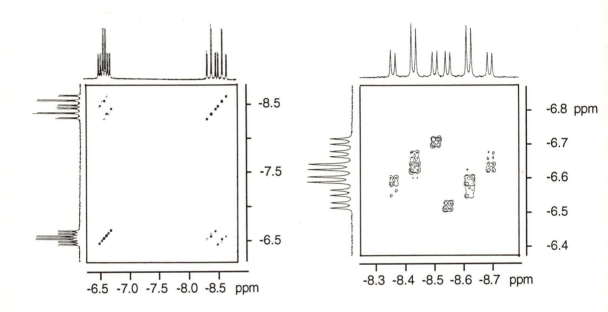

3.20 Vanadate ion and the ligand tricine form a complex in aqueous solution:

Interpret the two ^{13}C EXSY spectra on the following page. The solutions only differ significantly in their temperature. Signals from free ligand are marked **L**, and those from the complex are marked **C**.

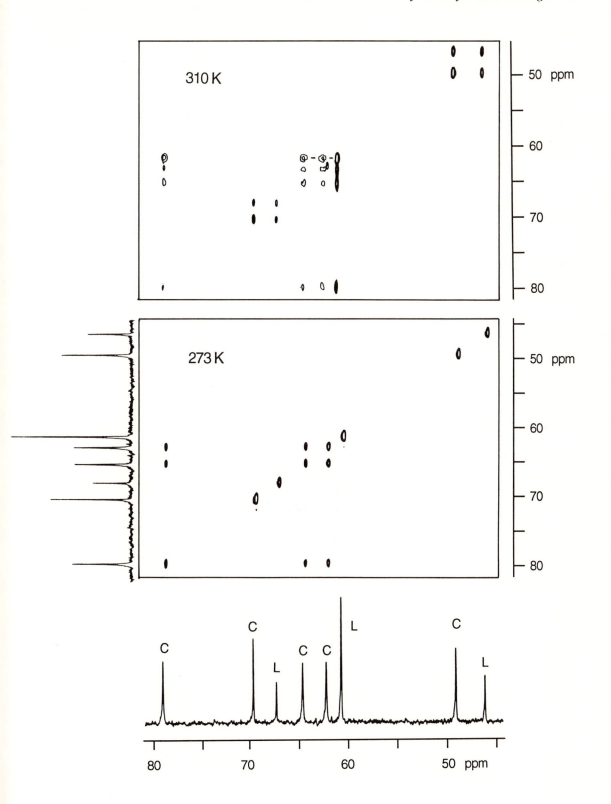

4 Structure determination using NMR alone

E4.1 The ^{51}V and ^{19}F spectra of the complex ion $[VOF_4]^-$ are shown here. What is the structure of the ion?

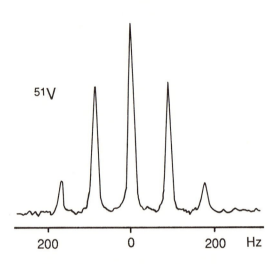

Answer The vanadium resonance is a well-resolved quintet. The splitting arises from four identical couplings to fluorine, which has $I = 1/2$. There is only a single fluorine resonance, showing that all four fluorines are chemically equivalent; the resonance is split into eight lines by coupling to ^{51}V, which has $I = 7/2$ and is 99.8% abundant. The apparent variation in height between these lines is an artifact arising from imperfect digitization. Clearly the ion has four-fold symmetry, and therefore it probably has the square-pyramidal shape shown.

It is not clear why the ^{51}V resonance should be so sharp: the ion cannot have the cubic symmetry that is normally required for slow relaxation of quadrupolar nuclei (§1.4.10). It is even more surprising that the relatively slow relaxation, and associated narrow linewidth, are retained in the mixed halide ions such as $[VOClF_3]^-$. This seems to be a common feature of vanadyl (V=O) ions, and presumably indicates a small electric field gradient in all these ions.

[Unpublished spectra described by Hibbert, R.C. (1985). *J. Chem. Soc. chem. commun.* 317–8.]

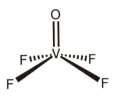

4.1 The reaction of $[nBu_4N][BF_4]$ with an excess of TaF_5 results in the formation of $[nBu_4N][Ta_2F_{11}]$. The ^{19}F spectrum of this product is shown below. What is the structure of the anion? Why do we not see any coupling to ^{181}Ta?

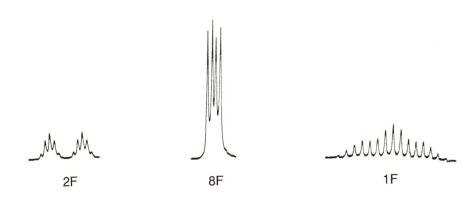

2F   8F   1F

4.2 A compound, $C_{12}H_{32}N_5^+$ Cl^-, newly isolated from bacteria, was only soluble in water. Both the ^1H and ^{13}C spectra, when acquired in D_2O, showed three signals of equal intensity. What is the structure?

^1H		^{13}C	
δ 2.2	qn	20.2	t
3.1	t	36.3	t
3.6	t	56.5	t

4.3 Use these low-temperature spectra to propose a structure for the complex cation $[CrH(CO)_2(Me_2PCH_2CH_2PMe_2)_2]^+$. The complete ^1H-decoupled ^{31}P spectrum is shown; only the hydride resonance (around -9 ppm) of the ^1H spectrum is shown.

At high temperatures, the hydride resonance collapses to a quintet, while the ^{31}P spectrum becomes a single broad line.

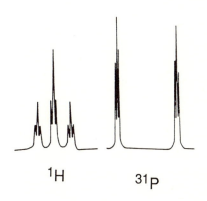

^1H ^{31}P

E4.2
Deduce the structure and relative stereo-chemistry of a compound, $C_6H_{12}O_6$, with the following spectroscopic properties. The 300 MHz ^1H spectrum, acquired in D_2O solution, is shown on the right, and summarized below. The 'missing' proton signals are in fast exchange with the solvent. The ^{13}C spectrum is also summarized below.

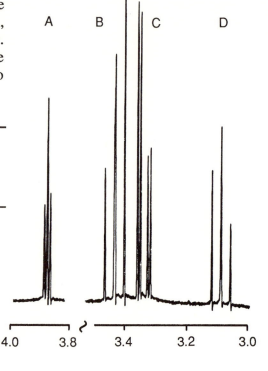

A B C D				

A	3.88	1H	t	2.8 Hz
B	3.46	2H	t	9.6
C	3.36	2H	dd	2.8, 9.6
D	3.10	1H	t	9.6

^{13}C spectrum

71.1	2C	d
72.2	1C	d
72.4	2C	d
74.3	1C	d

Answer The molecular formula indicates the presence of one double bond equivalent. The ^{13}C chemical shifts rule out the presence of unsaturation. The compound must therefore contain one ring.

Only six protons are carbon-bound, so the other six must be present in hydroxyl groups. Therefore all the oxygens are accounted for in the hydroxyl groups, and none are available for making rings; it follows that the ring contains only carbons.

All the carbons are CH, and their chemical shifts indicate that each is attached to a single oxygen. We can now rule out any branched chain rings because these would contain either quaternary and CH_2 groups (e.g. **a**) or carbons with two attached oxygens (e.g. **b**).

The compound must therefore be one of the stereoisomers of inositol (1,2,3,4,5,6-hexahydroxycyclohexane). There are eight possible isomers of inositol if enantiomeric structures are excluded. All eight isomers are shown opposite, the number inside the ring giving the number of carbon signals expected on the basis of symmetry.

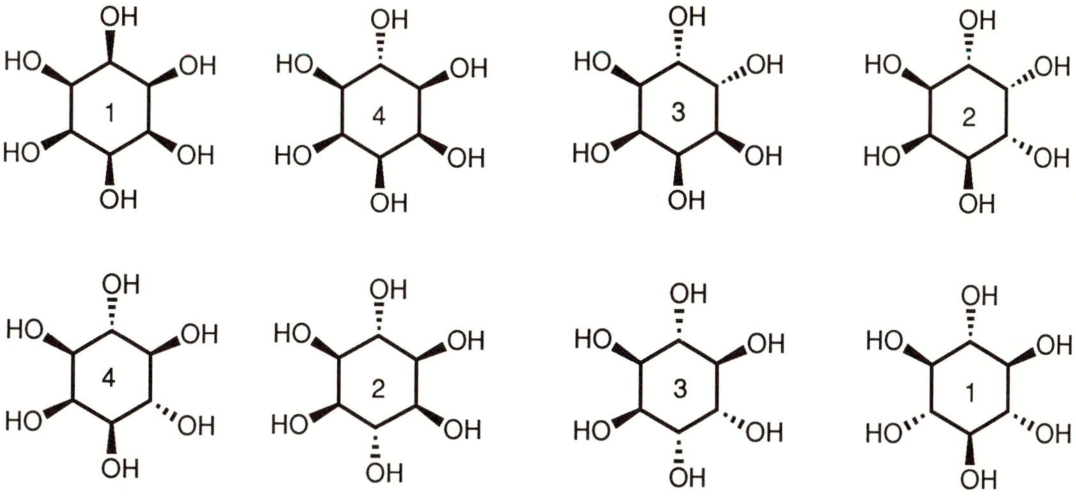

The observation of four carbon resonances now limits the possibilities to just two isomers; these can be distinguished by the couplings in the proton spectrum. Signal A is a 3 Hz triplet so it has two equatorial–equatorial or equatorial-axial couplings, while B and D are both 10 Hz triplets so they each have two axial–axial couplings. C experiences one axial–axial coupling and one small coupling. The only relative stereochemistry and conformation consistent with these is that shown to the right, *myo*-inositol.

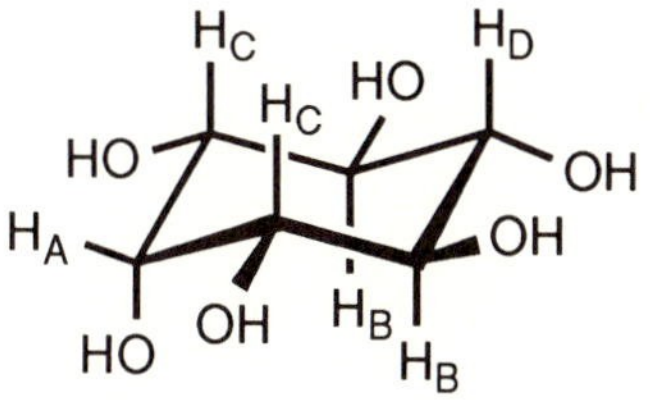

You should confirm that you understand the number of different carbon signals expected for each isomer; in some cases the symmetry elements are quite subtle. What pattern of proton coupling constants would you expect for the other isomer that has the correct symmetry?

4.4   How could you use ^{13}C and ^{1}H spectra to distinguish these isomers? Subtle knowledge of chemical shifts is not needed.

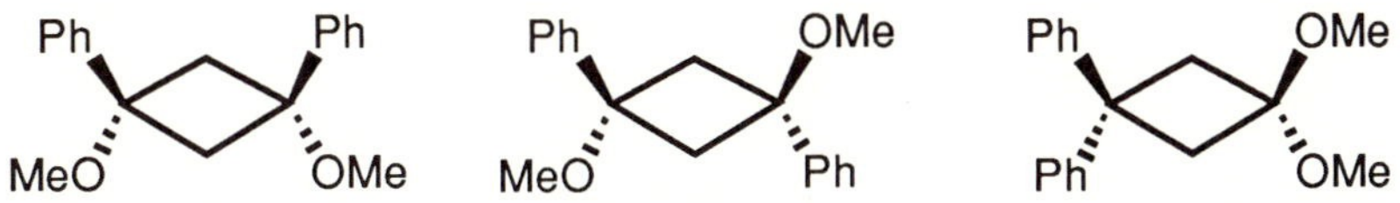

4.5 The two isomers of this amine show the following resonances in their ^{1}H NMR spectra. What are the isomers? Assign and explain their spectra.

Isomer (i)

2.74	3H	tt	3.9, 11.3 Hz
1.97	3H	td	3.9, 12.8
0.95	3H	td	11.3, 12.8

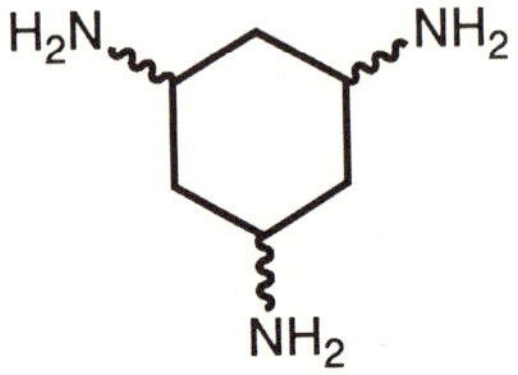

Isomer (ii)

3.42	1H	qn	3.8 Hz
3.13	2H	tt	4.1, 10.6
1.97	1H	ttd	1.9, 4.1, 12.3
1.71	2H	dddd	1.9, 3.8, 4.1, 13.5
1.21	2H	ddd	3.8, 10.6, 13.5
0.98	1H	td	10.6, 12.3

4.6 Methyl methacrylate, **a**, can be readily converted to the polymer shown, but the stereochemistry of the product depends on the reagents used: radical polymerization gives stereoisomer **b**, while anionic polymerization gives stereoisomer **c**. Deduce the stereochemistry of each isomer from the ^{1}H NMR properties of their CDCl$_3$ solutions.

b			c		
3H	s		3H	s	
3H	s		3H	s	
2H	s		1H	d	15 Hz
			1H	d	15

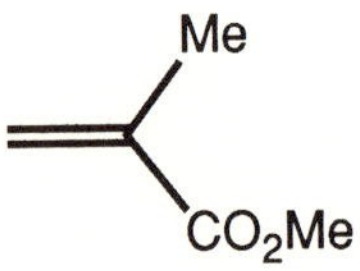

a

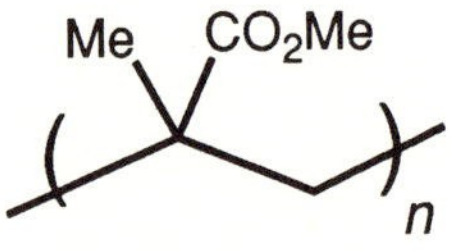

b, c

4.7   The ^{1}H-decoupled ^{13}C spectra below are of [Pt (η^3-C_3H_5)$_2$]
and the olefinic region of [Ni(η^3-1,1-Me$_2$C$_3$H$_3$)$_2$] in toluene-d_8
solution. What are the structures of these compounds?

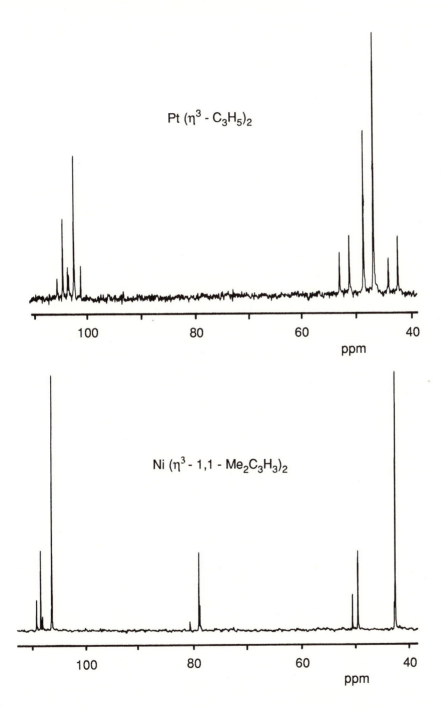

Pt (η^3 - C_3H_5)$_2$

Ni (η^3 - 1,1 - Me$_2$C$_3$H$_3$)$_2$

E4.3 The control spectrum **a** and NOE difference spectra **b** and **c** were obtained from a sample which was known to have one of the two structures shown below. The sample was dissolved in acetone and only the aromatic regions of the spectra are shown. Spectrum **b** results from pre-irradiation of the methoxyl protons, and spectrum **c** from pre-irradiation of the slowly-exchanging hydroxyl proton. On the basis of these spectra, which is the correct structure?

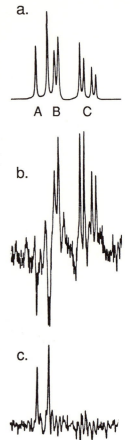

a.

b.

c.

A   B   C

a   **b**

Answer From the coupling patterns, the assignments of the signals must be as indicated on the two structures. The NOEs distinguish the structures on the basis of the proximity of the aromatic protons to the other parts of the molecule. Irradiation of the hydroxyl proton generates an NOE to A only, while irradiation of the methoxy group produces positive NOEs to B and C, with a negative NOE to A. This set of NOEs, which is illustrated below, is consistent only with structure **a**. Either of the difference spectra would have been sufficient to distinguish between the structures but it is reassuring to have the confirmation of the second.

Note that when the sample was dissolved in chloroform, the three aromatic signals overlapped so no analysis was possible. If the spectrum of any sample contains overlapping signals, it is always worthwhile to vary the solvent in the hope of obtaining a spectrum with more convenient chemical shift dispersion.

[Hunter, B.K., Russell, K.E., and Zaghloul, A.K. (1983). *Can. j. chem.*, **61**, 124–7]

4.8 The complex *trans*-[Pt(pyridine)$_4$Cl$_2$] [NO$_3$]$_2$ has been reported to decompose upon dissolution in water. The ^{195}Pt and 75 MHz ^{13}C spectra of an aqueous solution of the complex are shown below. What can you conclude about the major solution species? The ^{13}C chemical shifts of free pyridine in solution are 123.6, 135.5 and 149.7 ppm.

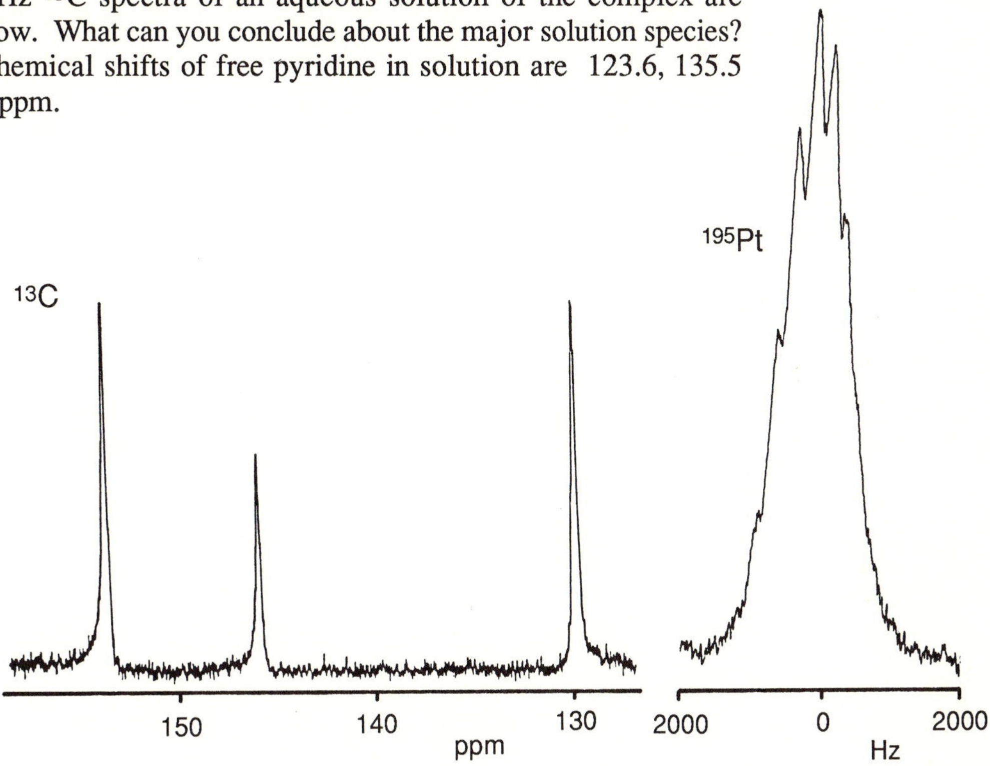

E4.4 A mixture of methyl ribosides **a** reacted with anhydrous acidic acetone to give a single dimer **b** with unknown stereochemistry at C$_1$. Use the NOE difference spectra on p. 76 to find the stereochemistry and conformation of **b**. Conditions: 360 MHz, CDCl$_3$ solution; two methyl singlets at 1.48 and 1.31 ppm are not shown.

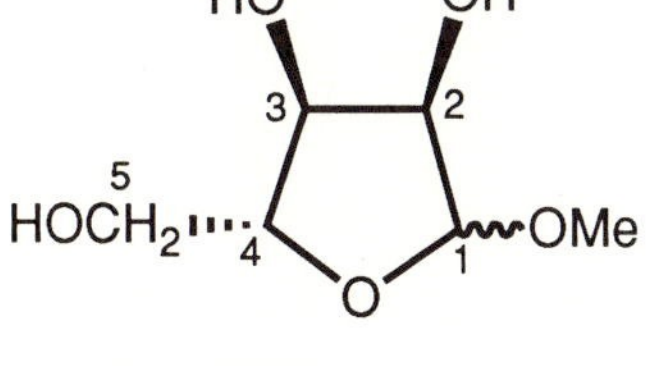

a 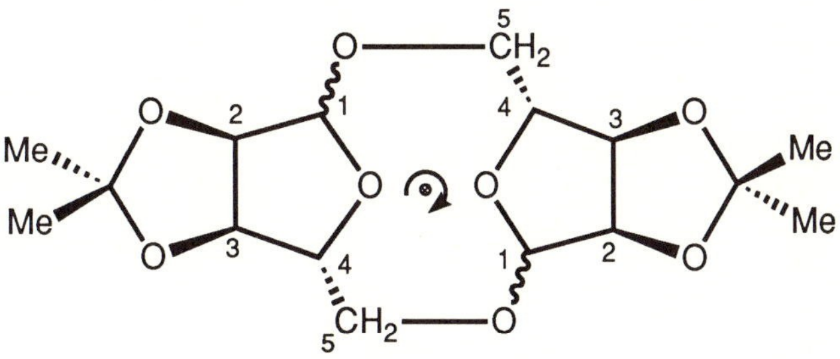   b

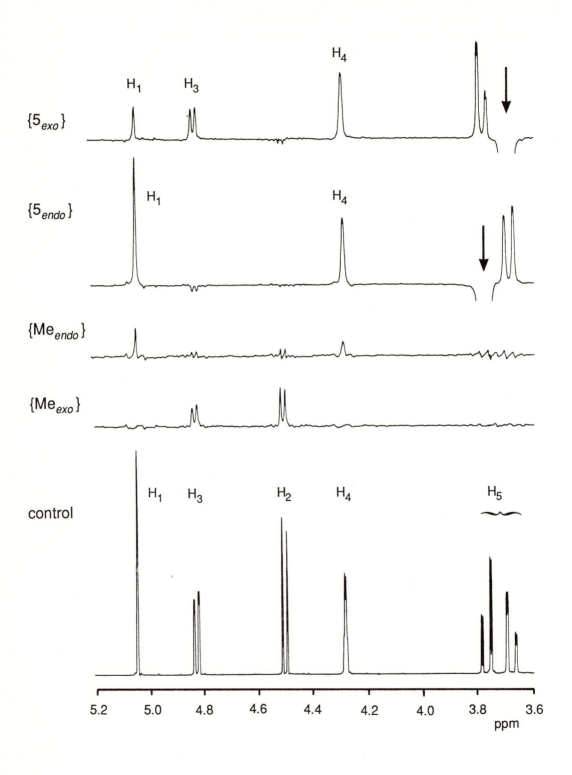

Answer The spectrum is so simple that the two ribose rings must have identical stereochemistry; there is a twofold axis of symmetry as shown in **b**. The assignments of H_{1-5} follow from decoupling experiments. Irradiation of one methyl singlet leads to large NOEs at H_2 and H_3; this must be the *exo*-methyl group (see **c**). The *endo*-methyl singlet gives NOEs to H_1 and H_4 defining the C_1-stereochemistry as shown in **d**.

c   **d**

Irradiation of the H_5 resonance at δ 3.77 gives large positive NOEs to H_1 and H_4 as well as to its geminal partner and a small, indirect, negative NOE to H_3. The other H_5 resonance, at δ 3.68, gives a much smaller NOE to H_1 but its NOE at H_3 is quite large. These results can be interpreted if the H_5 resonance at δ 3.77 is *endo*, while that at δ 3.68 is *exo*, and if the ten-membered ring has a major conformation as shown in **e**. Note that the NOEs from H_5 to H_1 are from one ribose group to the other, while the other NOEs are all within the same ribose moiety.

e

[Unpublished spectra described by Winkler, T. and Ernst, B. (1988). *Helvetica chim. acta*, **71**, 120–3.]

4.9 Deduce the stereochemistry and dominant conformation of this lactam. The ¹H spectrum shows the resonances listed below in addition to one exchangeable proton and five aromatic protons. Irradiation of C leads to NOEs at F, G and D.

A	3.5	1H	d	12 Hz
B	3.3	1H	dd	12, 2
C	2.6	1H	ddq	12, 6, 7
D	2.1	1H	ddd	13, 6, 2
E	1.8	1H	dd	13, 12
F	1.4	3H	s	
G	1.3	3H	d	7

4.10 Assign as far as possible the ¹H spectrum of this derivative of the anti-viral agent Virantmycin and determine the relative stereochemistry and conformation of ring B from the given NOEs.

					Enhanced by irradiation
A	7.7	1H	d	3 Hz	
B	7.6	1H	dd	3, 9	
C	6.5	1H	d	9	
D	4.4	1H	br s*		C
E	3.9	1H	dd	4, 6	M L K J H G
F	3.8	3H	s		
G	3.6	1H	d	10	H
H	3.5	1H	d	10	L G D
I	3.4	3H	s		
J	3.1	1H	dd	4, 16	N M K E A
K	2.8	1H	dd	6, 16	J E A
L	2.5	1H*	br s		K H G E
M	2.0	2H	m		N J H G E
N	1.8	2H	m		
O	1.6	9H	s		

4.11 A reaction product is known to have one of the two structures shown, which differ only in whether the carbonyl is 'south' and the epoxide is 'north' or *vice versa*. Use the spectra below to determine which is correct. Trace **a** is a ^{13}C spectrum acquired without any proton decoupling at any stage. Trace **b** is the ^{13}C NOE difference spectrum resulting from selective pre-irradiation of the non-aromatic single proton.

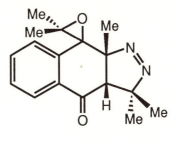

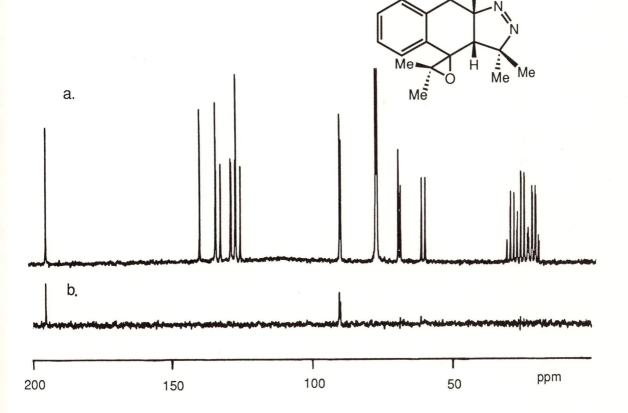

a.

b.

200   150   100   50   ppm

4.12 Deduce two possible structures and the conformation of this triphosphate from the following observations: (i) the ^{13}C spectrum shows six resolved signals; (ii) the ^{31}P spectrum shows three ^1H-coupled doublets, with J_{PH} of 4–8 Hz; (iii) the ^1H spectrum, taken in D_2O solution, has three accidentally coincident signals and three resolved signals: A (q, 8 Hz), B (dd, 3, 8 Hz), C (dd, 2, 3 Hz), where B is coupled to A and C; and (iv) A simplifies to a triplet when the ^{31}P resonances are irradiated, B and C being unaffected.

$R = 3 * H, 3 * PO_3{}^{2-}$

4.13 What is the structure of this compound, $Na^+ C_3H_6PO_5^-$, which was recently discovered in marine microorganisms? The spectra summarized below were acquired in D_2O solution.

The ^{31}P resonance was a ddd (6, 9, 12 Hz). The ^1H one-dimensional and COSY spectra are shown opposite and summarized below. Also given below are the ^{13}C spectroscopic details.

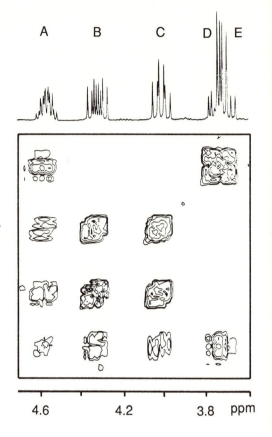

^{13}C

65	CH$_2$	
69	CH$_2$	
79	CH	

All three carbon signals show coupling to ^{31}P

^1H

A	4.57	dtt	4, 6, 7 Hz
B	4.32	ddd	6, 9, 12
C	4.02	dt	7, 9
D	3.76	dd	4, 12
E	3.70	dd	7, 12

4.14 A product from a photochemical dimerization is known to have the structure shown. Use the 3–5 ppm region of the ^1H spectrum opposite and the complete NOESY spectrum to determine fully the stereochemistry of this product. The NOESY spectrum was plotted without diagonal peaks.

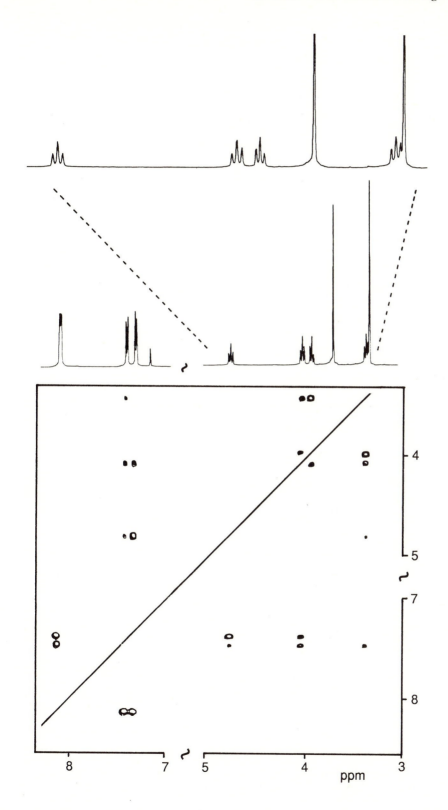

4.15 The phenol alboatrin, $C_{14}H_{18}O_3$, is a plant toxin found in a fungus. Determine its structure from the ^{13}C and ^1H spectroscopic properties summarized below.

^{13}C δ 16 q, 19 q, 22 t, 23 q, 36 d, 48 d, 74 t, 102 d, 107 s, 109 s, 110 d, 138 s, 154 s, 155 s.

Proton					Proton enhanced on irradiation
A	6.27	1H	d	2.4 Hz	
B	6.21	1H	d	2.4	
C	4.65	1H	br. s.*		
D	4.17	1H	dd	8, 8.5	
E	3.51	1H	dd	8, 8.5	D L
F	2.71	1H	dd	5.5, 17	
G	2.66	1H	dd	1.8, 17	
H	2.18	3H	s		
I	2.10	1H	m†		
J	1.93	1H	ddd	1.8, 5.5, 10.8	
K	1.51	3H	s		J
L	1.05	3H	d	7	E I I J

†Decoupling of I simplifies multiplets D, E, J, and L.

4.16 A sample of *t*-butyl lithium was prepared with isotopically pure ^6Li and ^1H-decoupled ^{13}C spectra were acquired over a range of temperatures. The traces show how the appearance of the quaternary carbon resonance depends on temperature. Use these spectra to propose a structure for the compound and to determine the nature of the fluxional process which is occurring.

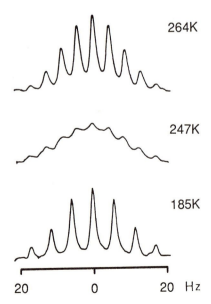

4.17 The anion $[B_9H_{14}]^-$ has the boron structural framework shown below; for clarity, none of the protons are shown. Metallation of this anion gives *nido*-$[(cp)RhB_9H_{13}]$. The proton-decoupled ^{11}B COSY spectrum of this metallated product is shown here. What is the structure of the product, and what information is missing?

(cp = cyclopentadienyl)

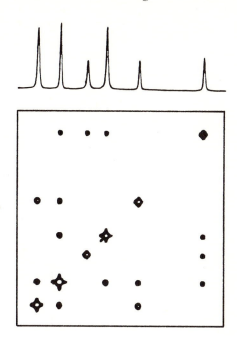

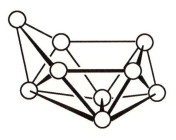

4.18 The hydride region of the 300 MHz ^1H spectrum of the octahedral complex $PtH_2(SnPh_3)_2(PMe_3)_2$ is shown here. The ^1H-decoupled ^{31}P spectrum contains a single resonance with a complex pattern of satellites indicating coupling to the central Pt ion and two equivalent tin nuclei. What is the structure of the complex?

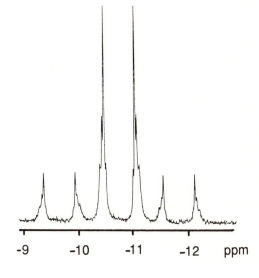

5 Structure and mechanism

E5.1 A newly isolated marine metabolite, $C_{11}H_9O_3NS$, has the following spectroscopic properties. What is its structure?

EI m/z 235 (M⁺), 151, 123, 112, 84.

v 3526, 1637 cm⁻¹.

¹H

A	8.3	1H	dd	1.9, 7.5 Hz
B	8.1	1H	d	3.8
C	8.0	1H	d	1.9
D	7.6	1H	d	3.8
E	7.0	1H	d	7.5
F	3.9	3H	s	

There is also one exchangeable proton.
Pre-irradiation of resonance F enhances only signal C.

Answer The molecular formula indicates the presence of eight double bond equivalents (DBEs). The splitting patterns for resonances A, C and E show the presence of a 1,2,4-trisubstituted aromatic ring, accounting for four DBEs; the infrared spectrum shows peaks for hydroxyl and conjugated carbonyl groups, taking a fifth DBE and using up two of the three oxygen atoms. The shift of F suggests a methoxyl group, accounting for the third oxygen. At this stage we have the four fragments shown below. Note that their nominal atomic weights add up to 151, a significant mass in the mass spectrum.

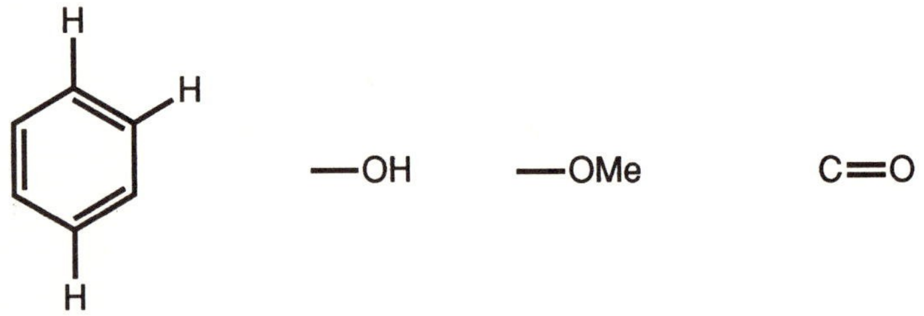

We have three remaining DBEs in the form of C_3H_2NS, which has a mass of 84. This number also appears in the mass spectrum, indicating that it probably corresponds to a stable fragment. Many structures are consistent with this formula, but the shifts of resonances B and D hint at a heterocyclic ring of the type and substitution pattern shown in **a**. The 3.8 Hz coupling is typical for vicinal coupling in five-membered aromatic rings. Furthermore, a proton on the carbon between N and S would have a shift of around 9 ppm.

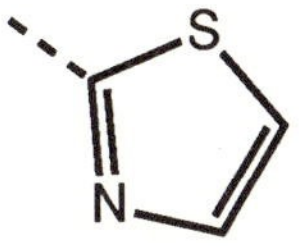

a 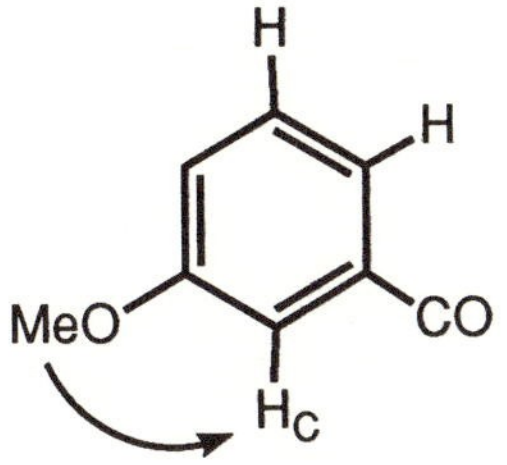   **b**

We now have to link all these fragments together: the chemical shifts of resonances A and C strongly suggest that they are adjacent to an electron-withdrawing group, which is presumably the carbonyl. Resonance C belongs to an isolated proton with only a *meta* coupling; it is enhanced by the methoxyl group, which must be adjacent, giving the arrangement shown in **b**. The only remaining substituent, the hydroxyl group, must go into the remaining position. This is confirmed by the deshielding of resonance E. Connecting the thiazole ring to the carbonyl group gives the final structure shown below, **c**.

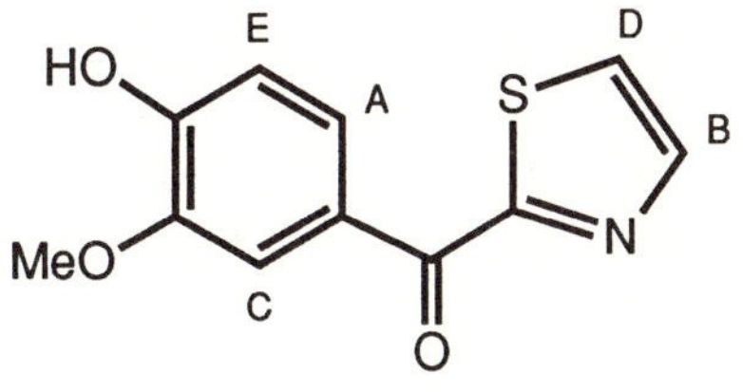

c 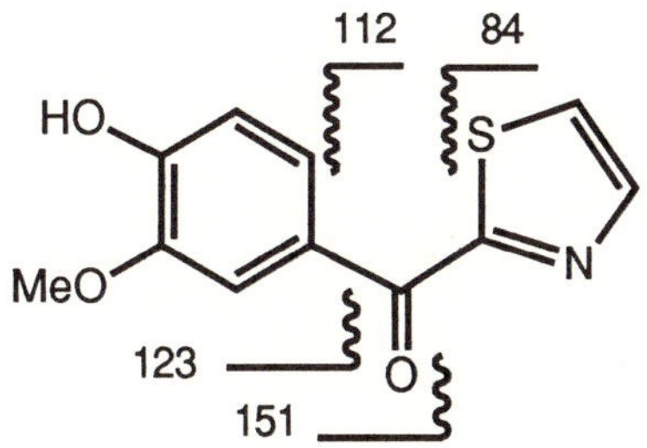   **d**

This structure is strongly supported by the mass spectrum: fragmentation on either side of the carbonyl group gives fragments of *m/z* 151 or 112, while loss of CO from each of these explains the peaks at 123 and 84. These fragmentations are summarized in **d**.

[Arabshahi, L. and Schmitz, F.J. (1988). *Tetrahedron letters*, **29**, 1099–102.]

5.1   Reaction of this pyridine derivative with acetone–anhydrous acid gave two isomeric products, $C_{11}H_{15}NO_3$. Routine ^1H spectra of each showed one 6H singlet, one 3H singlet, two 2H singlets and two 1H singlets. In dried DMSO solution, one compound gave an unchanged spectrum while the other showed mutual coupling ($J = 5$ Hz) of one of the 2H signals with one of the 1H resonances. What are the two products?

5.2   When this steroid reacts with MeMgBr in the presence of Cu^I, the product shows a 3H doublet ($J = 7$), at 0.9 ppm (^1H). Decoupling difference shows that the hidden multiplet to which this doublet is coupled is a ddq ($J = 2, 6, 7$ Hz). What is the product?

5.3   Assign the ^1H spectrum of the complex cation $[Ru(terpy)_2]^{2+}$ shown in trace **b** below. The spectrum was recorded in 1:1 DMSO-d_6/methanol-d_4. Resonances A and B relax faster than the other protons and A is coupled to C. Trace **a** shows the spectrum resulting one hour after the addition of an excess of NaOCD$_3$. What is occurring upon treatment with base?

Terpy

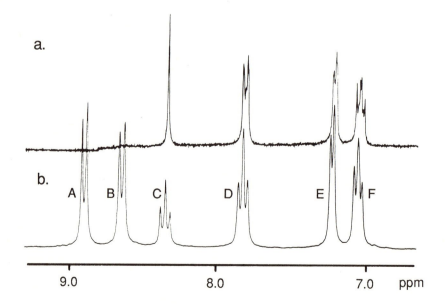

5.4 Reaction of this amine with ^{13}C-formaldehyde ($^{13}CH_2O$) and sodium cyanoborohydride ($NaBH_3CN$) in aqueous solution gave a solution with two signals in the ^{13}C spectrum: δ 41 q, 44 t. When the cyanoborohydride was replaced by NaCN, only the 44 ppm signal was seen. What are the products and how are they formed?

5.5 This compound reacts with hydroxide ion to give a product $C_{16}H_{12}O_3$. Deduce the structure of the product and suggest a mechanism for its formation.

ν 1700 cm^{-1}.

^{13}C: 12 signals for aromatic carbons (8 d, 4 s) and signals at δ 63 t, 74 t, 109 s and 189 s.

^{1}H

7–8	8H	m	
5.5	1H	d	16 Hz
5.3	1H	d	13
5.2	1H	d	13
4.9	1H	d	16

5.6 This bicyclic compound shows two resonances due to ring protons and two resonances due to ring carbons. On brief treatment with K_2CO_3 in methanol, it is partially converted to a new isomer, **i**; prolonged treatment under the same conditions yields a further five isomers, **ii-vi**. Identify the products.

Isomer	Number of ring resonances	
	^{1}H	^{13}C
i	6	6
ii, iii	3	3
iv	4	4
v	6	6
vi	2	2

5.7   Reaction of one equivalent of benzylamine ($PhCH_2NH_2$) with two equivalents each of formaldehyde (CH_2O) and phosphinic acid [$HP(OH)_2$] gives a product, $C_9H_{15}NO_4P_2$, with the spectroscopic properties summarized below. What is the product? The spectra were obtained in a D_2O solution.

The ^{13}C multiplicities below refer to ^{31}P-couplings; both of the carbons listed are methylenes. There are also four aromatic carbon signals.

^{31}P	10.0	2P	dt	$J_{PH} = 11, 554$ Hz
^{13}C	62.7	1C	t	$J_{PC} = 3.7$
	54.6	2C	dd	$J_{PC} = 3.9, 84$
^1H	8.0	5H	s	
	7.6	2H	d	$J_{PH} = 554$
	5.1	2H	s	
	3.9	4H	d	$J_{PH} = 11$

5.8   A terpene, newly isolated from the red alga *P. costatum*, is believed to have the structure shown. A 1H quartet (δ 5.9, $J = 1.5$ Hz) is coupled to a 3H signal at δ 1.9; saturation of the latter signal gives a 10% NOE to the 5.9 ppm resonance. What structural feature of the molecule does this experiment confirm?

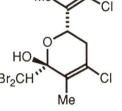

On standing in diffuse light at room temperature, the terpene is converted into a new compound with overall loss of water. Propose a structure for the new compound which is consistent with the following spectroscopic properties.

λ 272, 327 nm.

	^1H					^{13}C		
7.4	1H	d	14.0 Hz		188	s	142	s
6.8	1H	d	14.0		134	s	133	d
6.3	1H	s			130	s	125	d
6.2	1H	q	1.5		121	d	42	d
2.3	3H	s			19	q	18	q
1.9	3H	d	1.5					

5.9   Identify compound **L**, the product of this reaction sequence:

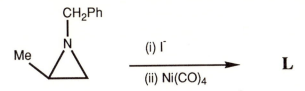

EI *m/z* 175, <u>133</u>, <u>132</u>, 105, 104, <u>91</u>; ν 1730 cm^{-1}.

	^{1}H			^{13}C	
7.2–7.4	5H	m		166.9	s
4.6	1H	d	15 Hz	47.0	t
4.1	1H	d	15	44.3	t
3.6	1H	m		44.1	d
3.1	1H	dd	5, 14	18.5	q
2.5	1H	dd	2, 14	+ 4 aromatics	
1.2	3H	d	7		

5.10   Fosfonochlorin, $C_2H_4O_4PCl$, is an antibiotic that was recently isolated from cultures of a soil fungus. It can take different structural forms, depending on the solvent and the pH, and is easily isolated as a sodium salt; it is stable in aqueous solution. Use the results below to define the contributing structures.

ν 3400, 1700, 1600 cm^{-1}.

λ (in acid solution) 276 nm; (in alkaline solution) 330 nm.

In D_2O the ^{1}H NMR spectrum contains only one signal (d, $J_{PH} = 5$ Hz).

The ^{13}C spectra in two different solvents are summarized below. Methine carbons are marked *, and methylene carbons are marked †; the remaining signals are quaternary:

DMSO		D$_2$O	
206	$J_{PC} = 175$ Hz	93	$J_{PC} = 198$ Hz
149	$J_{PC} = 201$	49†	$J_{PC} = 27$
103*	$J_{PC} = 47$		
51†	$J_{PC} = 74$		

E5.2
When ^{13}C-labelled formaldehyde, $^{13}CH_2O$, is fed to live cultures of bacteria in an NMR spectrometer, the metabolism of the label can be followed by ^{13}C NMR. Many bacterial species produce roughly equal amounts of formate (HCOO$^-$) and methanol (CH$_3$OH). This is reminiscent of the purely chemical Cannizzaro reaction in which a hydride ion (H$^-$) is transferred directly from one formaldehyde molecule to another. The accompanying 61 MHz deuterium NMR spectra are of methanol that results from the metabolism of deuterium-labelled formaldehyde, CD$_2$O, by *Escherichia coli* and *Pseudomonas putida*. What do they tell us about possible 'Cannizzarase' enzymes in those organisms?

P. putida

Answer A Cannizzarase operating on CD$_2$O would transfer a deuteride ion from one formaldehyde to another, generating CD$_3$OH:

$$2CD_2O + H_2O \rightarrow DCOO^- + CD_3OH$$

In water, the hydroxyl proton will be in fast exchange, so the deuterium spectrum of the product methanol would be a singlet. However, another possibility is that the oxidation and reduction of formaldehyde are independent processes carried out by separate enzymes. If this is the case, the outcome of the CD$_2$O feeding experiment will be different:

E. coli

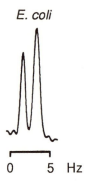

$$CD_2O + [O] + OH^- \rightarrow DCOO^- + HOD$$

$$CD_2O + [H^-] + H^+ \rightarrow CD_2HOH$$

In this case the deuterium spectrum of the product methanol will be a 1.7 Hz doublet by virtue of coupling to the geminal proton.

It is clear from the spectrum that *P. putida* metabolizes formaldehyde by some sort of Cannizzarase, as the bulk of the methanol produced is CD$_3$OH. The actual mechanism turns out to be quite unlike that of the chemical Cannizzaro reaction – see the reference below for details.

By contrast, *E. coli* produces almost exclusively CD$_2$HOH, showing that two independent pathways are operating. The difference in height of the two peaks is due to a small amount of CD$_3$OH, which is shifted upfield by an isotope shift effect of 0.02 ppm. At 61 MHz this is roughly half the size of J_{HD}. The 1.7 Hz splitting in CD$_2$HOH is removed by proton decoupling as expected. The small value of J_{HD} arises because the magnetogyric ratio of deuterium is only 1/6.55 of the value for protium; the measured 1.7 Hz corresponds to a geminal H-H coupling of around 11 Hz.

[Mason, R.P. and Sanders, J.K.M. (1989). *Biochemistry*, **28**, 2160–8.]

5.11 The bacterium *Nitromonas* biosynthesizes 3-nitropropanoic acid, **A**, from ammonia, NH_3. If the bacterium is grown on ^{15}N-labelled ammonia and natural-abundance oxygen gas ($^{16}O_2$), the ^1H-decoupled ^{15}N resonance of **A** is a sharp singlet; the shift of this singlet is arbitrarily defined as zero. When the experiment is repeated with ^{15}N-labelled ammonia and a 1:1 mixture of $^{18}O_2$ and $^{16}O_2$, **A** has the ^{15}N spectrum shown. What conclusions can be drawn about the source and mechanism of oxygen incorporation?

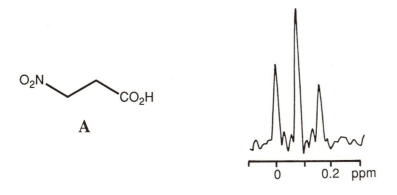

5.12 The ^1H spectrum of this compound in a non-polar solvent at low temperature contains the following signals:

A	6.0	1H	s
B	3.5	2H	ddd	4, 11, 14
C	2.7	2H	td	3, 11
D	1.5	2H	br. m

When the temperature is raised, or if a more polar solvent is used, resonance A is unchanged but B and C coalesce to give a 4H triplet.

If liquid sulfur dioxide (SO_2) is used as the solvent, the spectrum changes again: resonance A moves to 10.65 ppm, while B and C appear as a 4H triplet at 3.3 ppm. The ^{13}C spectrum in liquid SO_2 solution contains signals at 16, 31 and 185 ppm.

Explain these observations.

E5.3 Reaction of this penicillin-derived carbene, **a**, with thiophene, **b**, gives at least three adducts. The ¹H NMR properties of two of the adducts, summarized below, are consistent with addition of the carbene to the thiophene double bond. Use the observed NOEs to elucidate the stereochemistry of these adducts.

a

A third adduct results from expansion of the thiophene ring; proof of its structure and stereochemistry are given in §6.5.2.

b

					Group	Signals enhanced on irradiation
Isomer 1						
A	6.1	1H	dd	1.7, 6.0 Hz	T	B, C
B	5.8	1H	dd	3.0, 6.0	T	A, F
C	4.8	1H	s		P	-
D	4.4	1H	s		P	G
E	3.7	1H	dd	1.7, 7.2	T	F
F	3.2	1H	dd	3.0, 7.2	T	B, E
G	1.5	3H	s		P	B, D, F, H
H	1.1	3H	s		P	C, G
Isomer 2						
A	6.1	1H	dd	1.0, 5.4 Hz	T	B, C
B	5.7	1H	dd	3.0, 5.4	T	A, C, F
C	4.6	1H	s		P	A, B, H
D	4.2	1H	s		P	G
E	3.6	1H	dd	1.0, 7.3	T	A, F
F	3.4	1H	dd	3.0, 7.3	T	B, E
G	1.5	3H	s		P	D, H
H	1.1	3H	s		P	C, G

Answer The four possible stereoisomers **c**–**f** are shown opposite. They arise because the sulfur may be up or down and the ring may be to the left or right. The experimental shifts and couplings in the two adducts are essentially the same and provide no basis for distinguishing the possibilities.

c   d   e   f

g

The four singlet resonances belonging to the penicillin moiety are readily recognised. They are labelled P in the table opposite. The remaining four signals, labelled T, form a self-contained spin-system with mutual coupling. The assignments of resonances A, B, E and F to specific positions on the ring is straightforward on coupling and chemical shift grounds (see **g**). We can now use the observed NOEs to put the pieces of each molecule together like a jigsaw puzzle.

In isomer **1**, there are NOEs from the penicillin methyl resonance G to thiophene protons B and F, showing that the sulphur is down. An NOE at C from A indicates that the thiophene ring faces to the left, so isomer **1** is compound **d**.

In isomer **2**, there are no NOEs from the penicillin methyl groups G and H to the thiophene half of the molecule, indicating that the sulfur is probably up. There are NOEs from the lactam proton C to both olefinic protons on the thiophene half, indicating that the thiophene ring faces to the left, so this is compound **c**.

It is only the fact that we do observe NOEs from G to B and F in **1** that allows us to draw any safe conclusion from the lack of such NOEs in **2**. If **2** had been the only isomer available, the lack of NOEs would not have been a reliable guide to its structure.

[Mersh, J. D. and Sanders, J.K.M., unpublished results described by Chan, L. and Matlin, S.A. (1981). *Tetrahedron letters*, **22**, 4025–8.]

5.13 What is the structure of this metabolite, $C_{15}H_{18}O_2$, recently isolated from a fungal culture?

ν 1725, 1685, 1585 cm^{-1}; λ 260, 315 nm.

^{13}C

19.2 q	22.1 q	30.5 q	31.4 q	42.4 s
51.3 t	57.9 s	128.3 d	129.7 s	132.2 d
140.8 s	141.2 s	153.1 s	191.3 d	200.5 d

^1H

						Enhanced on irradiation
A	10.40	1H	s			
B	9.60	1H	s			
C	7.32	1H	d	8 Hz		
D	7.22	1H	d	8		
E	2.68	3H	s			A, D
F	2.21	1H	d	14		
G	1.86	1H	d	14		
H	1.55	3H	s			A, B, G
I	1.33	3H	s†			C, F, G
J	1.32	3H	s†			C, F, G

† Irradiated together

5.14 The ^1H spectrum of 1,3,5-trihydroxybenzene in water containing one equivalent of sodium hydroxide (NaOH) consists of a 3H singlet at 6.03 ppm. After addition of a second equivalent of NaOH, the spectrum shows two singlets (2H each) at 3.0 and 6.0 ppm. What are the compounds formed in these solutions?

5.15 Photolysis of the diazomalonate, **M**, in excess cyclohexane gives the adduct **N** via a carbene intermediate. Parts of the 250 MHz ^1H spectrum and *natural abundance* 38 MHz deuterium spectrum of **N** are shown opposite. What information do these spectra give about the mechanism of the reaction?

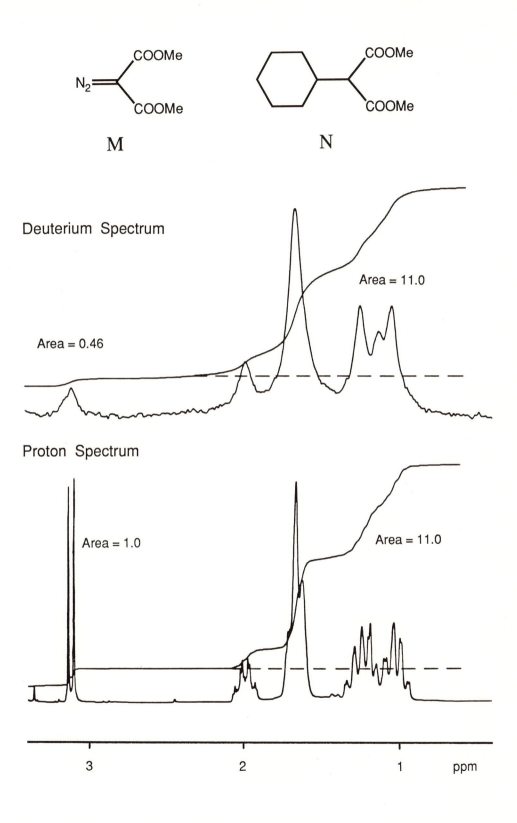

5.16 An optically-active insect pheromone, **P**, has the following spectroscopic properties. What is its structure?

EI *m/z* 154 (M$^+$, $C_{10}H_{18}O$), 136, 121, 87, <u>69</u>, <u>68</u>, <u>67</u>, <u>43</u>.

λ 226 nm; ν 3360, 3080, 1590, 985, 890 cm^{-1}.

^{13}C δ 0–50 (2 q, 2 t, 2 d), 100–140 (2 t, 1 d, 1 s).

^1H

A	6.4	1H	dd	12, 18 Hz
B–E	5.0 –5.3	4H	m	
F	3.8	1H	m	
G	2.5	1H	dd	4, 14
H	2.2	1H	dd	9, 14
I	1.7	1H	m	
J	1.6	1H	s*	
K	1.3	2H	m	
L	0.92	3H	d	7
M	0.90	3H	d	7

Decoupling of F simplifies G, H, and K.

P reacts with acetic anhydride to give a derivative of MW = 196. This derivative has a similar ^1H spectrum except that resonance F has moved to 5.1 ppm and resonance J has been replaced by a 3H singlet at 2.1 ppm.

5.17 Reaction of [HC≡NH]$^+$ [AsF$_6$]$^-$ with KrF$_2$ in HF solution at 210K led to immediate precipitation of an unstable solid, **A**. On warming, **A** decomposed violently to give, among other products, krypton gas but it was sufficiently stable in BrF$_5$ as solvent at 215K for the acquisition of spectra. A sample of **A** prepared with [HC≡^{15}NH]$^+$ gave ^1H, ^{15}N and ^{19}F spectra containing a single resonance each. These signals are summarized below. Propose a structure for **A**.

^1H	dd	4, 12 Hz
^{15}N	dd	12, 26
^{19}F	dd	4, 26

The ^{19}F resonance was complicated by the presence of satellites around ± 0.01 ppm from the main signal, and 20–30% of its intensity.

5.18 A toxic substance, **X**, $C_{11}H_{16}N_2O_5$, has recently been isolated from the Colorado potato beetle. Use the evidence below to deduce possible structures for **X**, which is soluble only in water.

FAB m/z 257 (MH)$^+$.

ν 3400-2500 (br), 1660, 1590 cm^{-1}.

λ 230 nm (ε 8000).

^{13}C

28.0 t	33.4 t	54.9 d	56.0 d	122.4 t
127.7 d	133.2 d	135.1 d	175.4 s	175.6 s
178.6 s				

^1H (D$_2$O solution)

6.75	1H	td	10.2, 16.7 Hz
6.26	1H	t	10.2
5.43	1H	dd	9.6, 10.2
5.38	1H	dd	1.8, 16.7
5.29	1H	dd	1.8, 10.2
5.06	1H	d	9.6
3.74	1H	t	6
2.45	2H	t	7.5
2.13	2H	m	

Hydrogenation of **X** with H$_2$/Pd/C gave **Y**:

CI *m/z* 261 (MH$^+$), 132, 130.

When trifluoroacetic acid is added to a D$_2$O solution of **Y**, a triplet at 3.8 ppm (1H, 6 Hz) shifts to 4.2 ppm.

Reaction of **Y** with 2,4-dinitrofluorobenzene, followed by acid hydrolysis (6N HC1, 12 hrs, 373K) gives one molar equivalent of **Z**.

Z

5.19 An antibiotic, **A**, has the formula $C_{14}H_{19}NO_4$. Deduce as much as you can about its structure and stereochemistry from the following spectroscopic data.

ν 3400, 3300, 1740 cm^{-1}.

EI *m/z* 144, 126, 122, 121, 84.

CI *m/z* 266, 206, 144, 121, 84.

The ^{13}C spectrum of **A** reveals the presence of two CH_3 signals, two CH_2 signals, five CH signals, and three quaternary signals. The only resonances with shifts greater than 80 ppm are a carbonyl and four aromatic carbons.

The full 300 MHz ^1H NMR spectrum of **A** in CDCl$_3$ solution is shown in trace **a** opposite; expanded multiplets from that spectrum are shown in trace **c**. The COSY spectrum, **d**, and the one-dimensional spectrum, **e**, on p. 100, were obtained in a CDCl$_3$–C$_6$D$_6$ mixture. Trace **c**, p. 99, shows expanded multiplets from the CDCl$_3$–C$_6$D$_6$ spectrum. Scale markers below the expanded multiplets are 10 Hz each.

All the proton signals are 1H each except A, B and L (2H each) and E and K (3H each). Resonances L are exchangeable with D$_2$O.

Three NOE experiments were carried out:

Irradiation of E enhanced B.
Irradiation of F enhanced C.
Irradiation of H/I enhanced A.

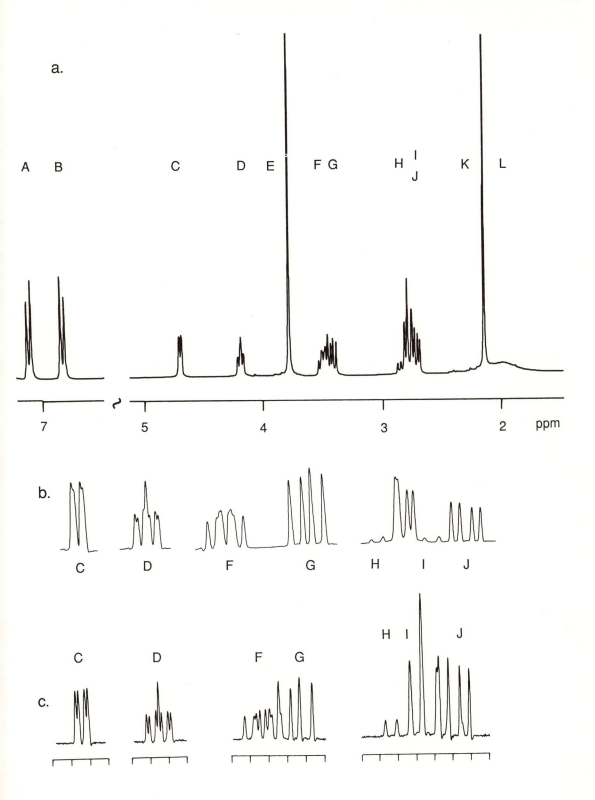

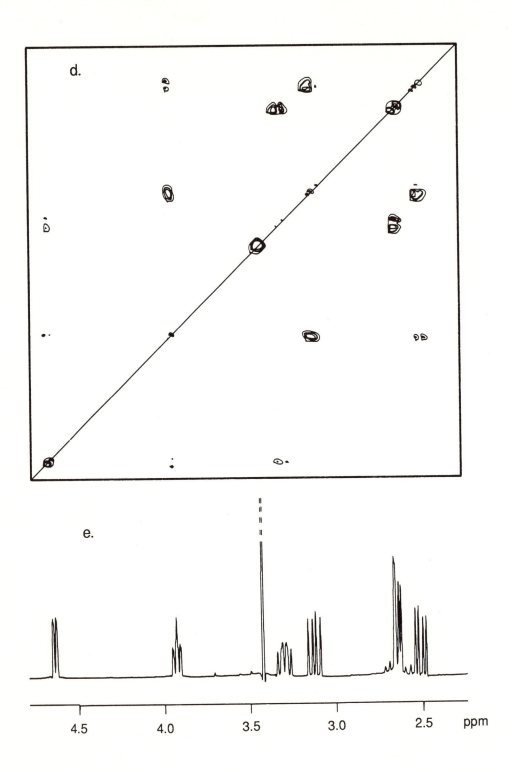

5.20 This two-dimensional ^{13}C exchange spectrum is of the methyl region of the heptamethylbenzene cation in concentrated sulfuric acid solution under conditions where methyl group migration is occurring. Is the migration process 1,2 or 1,3 or 1,4 or random?

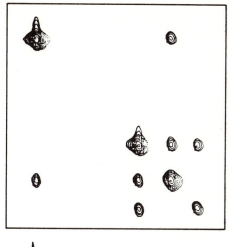

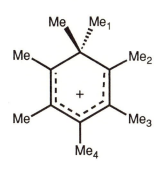

5.21 The *bis*-ferrocene, **A**, reacts with butyl-lithium to give a product, **B**, which reacts with methyl iodide to give **C**. The ^{13}C spectrum of **B** contains one resonance (δ 30, dd, $J_{CH} = 64, 130$ Hz) in addition to aromatic signals.

The ^1H spectrum of **B** at room temperature contains, in addition to aromatic signals, one resonance at δ 4.5 (1H, t, $J = 9$ Hz) and one at δ 3.1 (2H, d, $J = 9$ Hz). What is unusual about the structure of **B**?

A R = H
C R = Me

5.22 When a solution of **B** above is cooled to 170K, the 4.5 ppm signal changes to a doublet ($J = 18$ Hz) and the 3.1 ppm signal splits into two 1H signals at 3.5 and 2.7 ppm. What further insight does this give into the precise nature of **B**?

5.23 Deduce the structure of **A**, $C_{20}H_{22}N_2O_4$, a deeply coloured antibiotic with a UV spectrum similar to that of anthraquinone. **A** is the product of aerial oxidation of **B**, $C_{20}H_{24}N_2O_4$, which is colourless.

A reacts with methyl iodide, to yield three isomeric derivatives, **C–E**, $C_{22}H_{26}N_2O_4$. The ^1H spectra of **C** and **D**, which are summarized below, are very similar to that of **A**, apart from the 6H singlet, A, which has replaced a broad exchangeable 2H signal in the spectrum of **A**. The ^1H spectrum of **E** is more complex, containing a total of four 3H singlets.

	C	D			
A	3.7	4.2	6H	s	
B	3.0	3.1	4H	t	8 Hz
C	2.3	2.5	6H	s	
D	1.6	1.6	4H	m	
E	1.0	1.0	6H	t	7

The ^{13}C spectrum of **A** has 11 signals. The spectra of **C** and **D** are very similar to this, but both have 12 signals; the extra carbon is marked † in the table below. When protons A–C are irradiated during acquisition of the ^{13}C spectrum, some quaternary carbon signals sharpen due to loss of long-range coupling. These couplings are also indicated in the table below. Irradiation of signal A leads to a 60% NOE at the 179 ppm carbon of **C**, but not to the corresponding carbon of **D**. **E** has 22 carbon signals.

C		Coupled proton	D		Coupled proton
183	s		187	s	
179	s		182	s	
162	s	A, C	164	s	A, C
147	s	B, C	153	s	B, C
140	s	A	146	s	
134	s	B, C	126	s	B, C
118	s	B	126	s	B
34†	q		54†	q	
32	t		32	t	
23	t		23	t	
15	q		15	q	
14	q		12	q	

Hints

2.1 See p. 8 and §8.4.4.

2.2 See p. 13.

2.3 Predict the expected shift ranges and multiplicities from the structure, and use them to divide the carbon resonances into groups. Use the extra information about coupling to ^{1}H and ^{31}P to complete the assignment. Carbons 10 and 11 cannot be distinguished from these data.

2.4 (i) See p. 13. (ii) One-bond couplings are *very* large.

2.5 Do the chemical shifts of resonances A, B, F or G give you any obvious starting points for the assignments? How many axial and equatorial vicinal neighbours do F and G have?

2.6 Relaxation rates depend both on correlation time and inter-proton distances (§6.2). Look first at inter-proton distances in the substituted ring: it should be clear which protons will relax most rapidly. The relaxation rates of corresponding protons in the two rings will tell you about the difference in effective correlation time.

2.8 (i) ^{13}C-labelling does not change the intensities of these satellites. (ii) See p. 13.

2.9 First look at the COSY spectrum to sort the multiplets into coupled partners; one signal in each pair can be assigned on chemical shift grounds, using the electronic effects of the substituents. Remember that the centre lines of triplets are missing from COSY cross-peaks (§4.3.1). Then find the extra NOESY connections in the spectrum. Finally, look at the structure and predict which protons should be connected by NOEs.

2.10 Assign the ^{1}H spectrum first, using the coupling patterns and chemical shifts.

2.11 How far are the various carbons from the nearest protons?

2.12 One signal is missing from the C–H correlation because of an unusually large C–H coupling constant.

2.13 How many types of pyridine ring are there? Look at a model to predict possible ring current effects.

2.14 How do heteronuclear couplings behave in a homonuclear *J*-resolved spectrum? See §8.2.2.

2.15 How does solvent viscosity depend on temperature? How can that affect signal intensity in a pulsed experiment?

2.16 Start from secure proton assignments such as the ethyl methyl group: there can be no two-bond correlations to a quaternary carbon, and the only three-bond correlation must be with C_3. In this way, you should be able to find the following correlations in the spectrum:

Two bond correlations   Three bond correlations

2.17 The negative NOEs result from an indirect 'three-spin' effect involving fluorine as the central spin (§6.2.3).

2.18 The COSY spectrum contains several two- and four-spin systems which are partly connected by faint long-range correlations. To obtain complete assignments, it will probably be necessary for you to move several times between the COSY plot, the NOE results and a good model or three-dimensional diagram; good starting points are G, P, Q, and R. For guidance on the NOEs expected in this type of system see §6.5.1. or Hall, L.D. and Sanders, J.K.M. (1981), *J. org. chem.*, **46**, 1132–8. The latter paper also describes some long-range couplings that are useful for assignments in this type of molecule.

2.19 The shift of one carbon is so characteristic that it provides an unambiguous starting point for analysing the INADEQUATE spectrum. The ^{1}H assignments follow from the C–H correlation. However, it is not practical to distinguish axial from equatorial protons from the small scale spectrum shown here.

3.1 What is the nitrogen hybridization? Look carefully at the spatial arrangement of the ring protons.

3.3 How many different ^1H signals do you expect? Is the exchange process intra- or intermolecular? The ^{11}B spectrum contains nine lines.

3.4 Which different isotopic species are formed? Are they in exchange with each other? What coupling pattern would each show? Are there any isotope effects on the shifts? See §7.2.3.

3.5 Can you see any stereochemistry in the molecule? What is the effect of an acid on an amine? See §7.2.3.

3.6 Are we detecting a coupling or a slow exchange process?

3.7 Why are there two methyl signals, and what is the simplest way for them to interconvert? What type of reaction increases in rate with concentration?

3.8 (i) See §7.2.3. (ii) Use the protons that change on acidification as starting points. (iii) Predict the splitting patterns for each proton around the ring.

3.9 See §7.2.2.

3.10 Make a model or careful drawing. Is the molecule planar?

3.11 The signal doubling arises because the two enantiomers of **b** are in different environments as a result of an interaction with **a**.

3.12 What is the apparent symmetry of the porphyrin in solution and in the crystal? Also, see §9.9.

3.13 Are the protons of the 'cyclohexatriene' group exchanging directly with each other?

3.14 At high temperature or in the presence of acid there is an exchange process going on. What is it? Is it possible that at lower temperatures the exchange is affecting relaxation rates but not shifts? See §7.2.5 and §7.3.3.

3.15 How many chiral centres are there? See Appendix.

3.16 The bonds about the nitrogen are planar. What is the spatial relationship between the aromatic ring and the remaining nitrogen substituents?

3.17 What is the effect of amide rotation on the aromatic protons?

3.18 Consider the effects of the concentrations and the removal of acid (and some water) by the alumina. See §7.2.6.

3.19 How would you label each proton line in terms of the α and β labels of the phosphorus atoms and the other proton? How does the exchange process change these labels?

3.20 Are all the CH_2OH 'arms' equivalent in the complex? Can they be interconverted intramolecularly?

4.1 The two Ta centres are equivalent and octahedral.

4.2 If the ^{15}N spectrum had been acquired without NOE under conditions which allowed accurate integrations, it would have shown two signals in the intensity ratio 1:4.

4.3 (i) The splittings in the ^{31}P signals are P–P couplings. (ii) The splittings in the ^1H signal are H–P couplings.

4.4 What are the similarities and differences in symmetry between the three molecules? Look for planes and axes of symmetry.

4.5 (i) What are the possible isomers? (ii) Don't forget about the possibility of four-bond 'W'-coupling.

4.6 Think about the symmetry of the environment created by the substituted carbons on each side of the methylene group.

4.7 (i) Consider the orientation of the π-allyl groups. How many isomers are possible in each case? (ii) Can the metal ion affect the spectrum? See p. 13.

4.8 Is the size of the observed ^{195}Pt–^{14}N coupling compatible with the ^{13}C spectrum of the pyridine being a fast exchange average between free and bound? See §7.2.3.

4.10 The key assignments to make are the hydroxyl proton and the methylene protons in the CH_2OMe group. Do you expect this methylene pair to be equivalent?

4.12 Use the ^{13}C spectrum to establish the symmetry of the molecule, and the ^1H spectrum to establish as much as possible about which groups are axial or equatorial.

4.13 Hydrolysis of the compound gives glycerol as the sole organic product.

4.14 (i) The effective shape of a cyclobutane is as shown. (ii) Assign the signals as aromatic, ring methine or ester methyl, and determine the symmetry of the product. (iii) Write down all the possible structures and predict the NOE connections to find out which groups are on the same face of the ring.

4.15 (i) A coupling constant of 17 Hz can only occur in a CH_2 group adjacent to a π-system. (ii) The combined effect of two electronegative groups attached to a single carbon can bring that resonance into the 'aromatic' region of the spectrum.

4.16 How many lithium atoms are coupled to the carbon at low temperature and at high temperature? See p. 13.

4.17 For a good discussion of this type of chemistry see Greenwood, N.N. and Earnshaw, A. (1984), *Chemistry of the elements*, Pergamon, Oxford, Chapter 6.

5.1 Both products are acetals (ketals).

5.2 The Grignard reagent can add either directly to the carbonyl group, or at the end of the conjugated system; it can also attack from above or below. Predict the spectrum, particularly the multiplicities and couplings, for all the possible products.

5.3 Resonances A and B are not coupled but they relax each other efficiently because they are close in space.

5.4 (i) Amines react with formaldehyde to give iminium ions, $(R_2N=CH_2)^+$ which can be attacked by nucleophiles. (ii) In this spectrum only the labelled carbons are observed.

5.5 Is the molecule still planar? Use the molecular formula and the ^{13}C spectrum to deduce how many new rings have been formed.

5.6 Think about reversible formation and protonation of enolate anions, and about the symmetry of the resulting isomers. The bicyclic ring structure is unchanged in all the products.

5.7 Amines react with formaldehyde to give iminium ions, $(R_2N=CH_2)^+$ which can be attacked by nucleophiles.

5.10 (i) Think about keto and enol forms. (ii) Carbonyl compounds with electronegative substituents often form stable hydrates; the ^{13}C shift of formaldehyde hydrate $[CH_2(OH)_2]$ is around 83 ppm.

5.11 (i) Both ^{16}O and ^{18}O have $I = 0$. (ii) How does the distribution of isotopes depend on whether one or both oxygens of the NO_2 group are derived from O_2? (iii) If both are derived from O_2, how does the distribution depend on whether they are from the same molecule of O_2?

5.12 (i) Conformational equilibria are not involved. (ii) Liquid SO_2 is a very polar, non-nucleophilic, solvent. (iii) Addition of a nucleophile such as PhSH gives a good yield of this substitution product.

5.14 (i) Only carbon-bound protons are visible. (ii) The products have lost one and two hydroxyl protons respectively. (iii) Are there any other tautomeric forms available?

5.15 (i) Assign the signal at 3.1 ppm. (ii) The integrals in both spectra are reliable. (iii) Why should there be less than the average amount of deuterium in the position giving a signal at 3.1 ppm?

5.17 The major isotopes of krypton, with their abundances, are ^{82}Kr (12%), ^{83}Kr (12%), ^{84}Kr (57%), and ^{86}Kr (17%).

5.19 The exchangeable protons are from an NH and an OH group.

5.20 Migrate one methyl group from one position to another, and see how the label (1,2,3,4) for each carbon changes. That gives you the appropriate cross-peak. Repeat the process for each of the four mechanisms, and see which fits the experiment.

5.21 You can assume that spin-spin coupling is a through-bond phenomenon. The observation of a half-sized C-H coupling is particularly significant.

5.22 The room temperature spectrum in the previous problem is a fast exchange average between how many structures? Are they the same or different? Are they symmetrical? What is the shape of the potential well?

5.23 (i) What is the symmetry of **A**? (ii) An amide group, O=C-NH, can be alkylated either on oxygen or on nitrogen.

Solutions

2.1 'CD$_2$Cl$_2$' usually contains about 99.8% deuterium. The great majority of the residual protons will be found in CHDCl$_2$ molecules, and will therefore resonate as a 1:1:1 triplet by virtue of coupling to deuterium. The measured HD coupling, 1.1 Hz, corresponds to an HH coupling of 7.2 Hz, because $J_{HH} = 6.55J_{HD}$. The small number of protons present as CH$_2$Cl$_2$ resonate as the downfield singlet. The isotope effect on the chemical shift is 0.012 ppm.

2.2 The isotopes ^{47}Ti and ^{49}Ti have such similar magnetogyric ratios that they are easily observed in the same spectrum. Each gives only a single resonance for TiCl$_4$.

2.3

Carbon	Shift	Carbon	Shift	Carbon	Shift
1	41.3	2	207.3	3	61.0
5	68.5	6	51.8	7	173.2
8	65.8	9	21.9	10,11	69.0, 69.3

2.4 Grossel, M.C., Moulding, R.P., and Seddon, K.R. (1983). *J. organomet. chem.*, **253**, C50–4.

2.5 The splittings (in Hz) measured from the spectrum are summarized in the table. Three large couplings to F show that it has two axial neighbours as well as a geminal partner, and so the methyl group must be equatorial, as expected. An unresolved 'W-rule' coupling to C distinguishes the broad equatorial A from the much sharper axial B.

	A	B	C	D	E	F	G	H
A		6	~1					
B	6							
C	~1			11	5	~1		
D					11	~3	6	
E			11			12	~2	
F			5	11	12		13	
G			~1	~3	~2	13		
H				6				

2.6   The protons in the unsubstituted ring relax slowly because ring rotation is fast.   In the substituted ring, protons with one neighbour relax more slowly than those with two neighbours. Adapted from Adam, M.J. and Hall, L.D. (1980).   *J. organomet. chem.*, **186**, 289–96.

2.7   Lycka, A., Jirman, J., Nobilis, M., Kvasnicková, E., and Hais, I.M. (1987). *Magn. reson. chem.*, **25**, 1054–7.

2.8   Holmgren, J.S., Shapley, J.R., and Belmonte, P.A. (1985). *J. organomet. chem.*, **284**, C5–8.

2.9   [Smith, W.B., unpublished results.]

2.10   [Wray, V., unpublished spectra.]

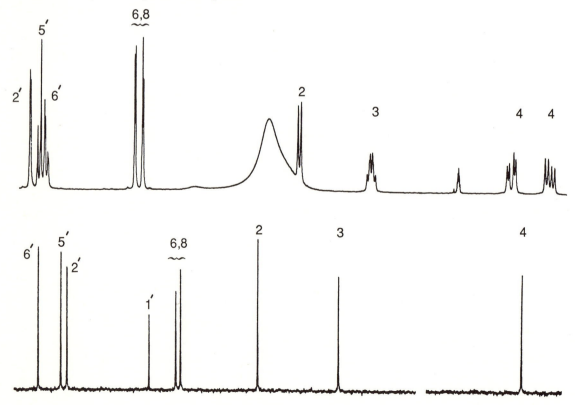

2.11 The methine carbons carry protons so they have full NOEs and relax rapidly. They give the intense signals at 110 and 134 ppm. Three carbons have no directly-attached protons but are quite close to protons because they are attached either to NH_2 or to CH_3 groups. The unique carbon at 147 ppm is attached to the nitro group; it is most distant from protons, so it relaxes the slowest and gives the weakest signal.

2.12 Guella, G., and Pietra, F. (1991). *Helv. chim. acta*, **74**, 47–54.

2.13 The two rings of each 2,2'-bipyridyl are inequivalent, and the connectivity for each ring is derived from the COSY spectrum. Binding to the electronegative Ir ion generally deshields the bipyridyl protons and increases their chemical shifts. Models reveal that H_6 of ring B lies over the face of ring A of the other bipyridyl group; it is shielded by ring A, and so appears at 8.1 ppm. By contrast, H_6 of ring A lies close to the edge of ring B and is shifted even further down-field than the other protons to 9.8 ppm.

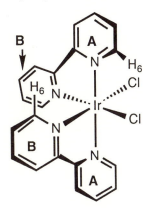

[Constable, E.C. and Leese, T.A., unpublished results.]

2.14 A = $H_{3,5}$(ax); B = $H_{3,5}$(eq); C = H_4(ax); D = H_4(eq). The couplings, in Hz, are summarized in the table. The large coupling of B to phosphorus results from a 180° dihedral angle between the spins.

Proton	A	B	C	D
A	-	12	12	2
B	12	-	5	2
C	12	5	-	15
D	2	2	15	-
^{31}P	4	22	1	2

[Pearce, C.M. and Sanders, J.K.M., unpublished results.]

2.15 Signals are weak at high temperature because there is insufficient relaxation time between pulses. Stronger signals in the low-temperature spectrum are a consequence of the more rapid relaxation resulting from a longer τ_c in the more viscous solvent. Some signals show only a small intensity difference in the two spectra; these belong to the nuclei that relax most rapidly. Using a smaller flip angle would reduce the saturation at high temperature.

[Chaloner, P.A. and Sanders, J.K.M., unpublished spectra.]

2.16 Cavanagh, J., Hunter, C.A., Jones, D.N.M., Keeler, J.H., and Sanders, J.K.M. (1988). *Magn. reson. chem.*, **26**, 867–75.

2.17 Sánchez-Ferrando, F. and Sanders, J.K.M. (1987). *Magn. reson. chem.*, **25**, 539–43.

2.18 A and B are coincident; Q and R cannot be reliably distinguished as we are not sure of the preferred conformation of the *iso*-propyl group. Note the small but characteristic COSY cross peak between the angular methyl P and 1α. The 13'- and 14'-hydroxyl protons are in slow chemical exchange, but are exchanging rapidly enough with water in the solvent to show saturation transfer.

Signal	A,B	C	D	E	F	G	H	I
Proton	16	12	11	7	14	14'	18	5

Signal	J	K	L	M	N	O	P	Q,R	S
Proton	2α	2β	6α	6β	1β	1α	15	19,20	13'

[Amor, S.R., Matlin, S.A., Pearce, C.M. and Sanders, J.K.M. (1989). To be submitted.]

2.19 The obvious starting point is the resonance at 71 ppm, which must belong to the carbon bearing the hydroxyl group. The remaining assignments follow from the INADEQUATE spectrum, apart from the *iso*-propyl methyl carbons which cannot be distinguished here.

3.1   Adapted from   Dutler, R., Rauk, A., and Sorensen, T.S. (1987). *J. Am. Chem. Soc.*, **109**, 3224–8.

3.2   Jochims, J.C., von Voithenberg, H., and Wegner, G. (1978). *Chem. ber.*, **111**, 1693–1708.

3.3   All of the protons and all of the borons are equivalent on the chemical shift time-scale. The boron couples to eight equivalent protons to give a nine-line multiplet and the protons couple to three equivalent ^{11}B nuclei to give a ten-line (1:3:6:10:12:12:10:6:3:1) multiplet. The exchange must be intramolecular or all coupling would be lost. Roughly half of the protons are broadened by attachment to one or more ^{10}B nuclei, giving rise to the broad hump.

3.4   Sanders, J.K.M., Hunter, B.K., Jameson, C.J. and Romeo, G. (1988). *Chem. phys. letters*, **143**, 471–6.

3.5   Adapted from Saunders, M. and Yamada, M. (1963). *J. Am. Chem. Soc.*, **85**, 1882–4.

3.6   Feeney, J., Partington, P., and Roberts, G.C.K. (1974). *J. magn. reson.*, **13**, 268–74.

3.7   Song, S.K., Watkins, C.L., and Krannich, L.K. (1987). *Magn. reson. chem.*, **25**, 484–8.

3.8   Resonances A and B are slowly-exchanging hydroxyl protons; they exchange on addition of acid, and this decouples them from C and G. C, the carbon-bound proton with the largest chemical shift, is likely to be on the anomeric carbon with two attached oxygens. D and E are distinguished by the number of couplings. F and H may be reversed.

3.9   Rotation of the NMe$_2$ group leads to exchange of the two methyl environments; this occurs at a rate, k, comparable with the NMR chemical shift timescale. The coalescence temperature depends on the frequency difference Δv (in Hz, *not* ppm) between the two signals (§7.2.1). At coalescence, $k = 2.22\,\Delta v$. The 60 MHz spectrum is at coalescence; the frequency difference is only 8.5 Hz. At 400 MHz, the frequency difference is 57 Hz and the methyl resonances are close to slow exchange. The 200 MHz spectrum is in intermediate exchange.

3.10 The planar conformation is highly strained so the ring is puckered, making the two protons of each CH_2 group inequivalent; one is in a pseudo-equatorial position, and the other pseudo-axial. Interconversion through the planar transition state is slow on the chemical shift timescale.

3.11 Compounds **a** and **b** form two diastereoisomeric ion pairs. Parker, D. and Taylor, R.J. (1987). *Tetrahedron*, **22**, 5451–6.

3.12 The molecule crystallizes with successive porphyrins stacked in an offset geometry that reduces the effective symmetry from fourfold to twofold. The large chemical shifts arise from an intermolecular ring current effect of the porphyrin π-system. Other common reasons for the doubling of signals in the solid are the presence of two non-equivalent molecules in the unit cell and the slowing of conformational processes that are rapid in solution. [Anderson, C.J. and Hunter, B.K., unpublished spectra; see also Leighton, P., Cowan, J.A., Abraham, R.J., and Sanders, J.K.M. (1988). *J. org. chem.*, **53**, 733–40.]

3.13 Bandy, J.A., Green, M.L.H., and O'Hare, D. (1986). *J. Chem. Soc., Dalton trans.*, 2477–84.

3.14 Vander Elst, L., Van Haverbeke, Y., Maquestiau, A., and Muller, R.N. (1987). *Magn. reson. chem.*, **25**, 16–20.

3.15 There are two chiral centres, generating meso and racemic isomers. The methylene protons in each ethyl group are inequivalent and the pattern at 2.1 ppm is the AB parts of two overlapping ABX_3 spin systems.

3.16 **a**: Rapid rotations occur so the spectrum shows the expected triplet and quartet for the ethyl group and singlets for the methylene and methyl groups. **b**: Rotation about the N–aromatic bond is slow and, with the aromatic ring perpendicular to the page, the molecule has a left and a right side so that, even with rapid rotation of the ethyl groups, the methylene protons in each ethyl group are inequivalent. Try the substitution test (Appendix) on the methylene protons in the heterocyclic ring and in the ethyl group. **c**: There are two rotational isomers about the CO–N bond and they are in slow exchange. There are two sets of signals of unequal intensity each displaying the same properties as **b**.

d: Both the X pair and the Y pair show inequivalence. Therefore, the molecule must have a top and a bottom as well as a left and a right side; the aromatic ring must be in slow rotation. and either there is only one amide isomer or the isomers are in rapid exchange. The X and Y pairs are AE and CD but we cannot say which pair is which. The ethyl methylene protons accidentally have the same chemical shift.

3.17 The chemical shift of proton A in the two rotamers is very different because of the adjacent carbonyl group. The remaining protons are largely unaffected. The rate of exchange is fast on the chemical shift timescale for these protons but intermediate for proton A where the chemical shift difference is much larger. Even though the A signal is exchange broadened, you still observe the coupling to it because it remains attached to the same carbon.

3.18 (i) In very dilute solutions all of the hydroxyl groups are in slow exchange and show coupling to the neighbouring A protons. (ii) The hydroxyl protons are now rapidly exchanging and decoupled. The exchange must be between one alcohol molecule and another (not with water) as it is dependent on diol concentration. (iii) The hydroxyl protons in propanediol are intramolecularly hydrogen bonded as shown, so they exchange more slowly with other molecules: this also explains the chemical shift difference between the hydroxyl resonances in (a) and (c).

3.19 Ball, G.E. and Mann, B.E. (1992). *J. Chem. Soc. chem. commun.*, 561–3.

3.20 Crans, D.C., Ehde, P. M., Shin P. K., and Pettersson L. (1991). *J. Am. Chem. Soc.*, **113**, 3728–36.

4.1 Brownstein, S. (1973). *Inorg. chem.*, **12**, 584–9. The Ta relaxes too quickly for any coupling to be resolved.

4.2 $[N(CH_2CH_2CH_2NH_2)_4]^+ Cl^-$.

4.3 Salt, J.E., Wilkinson, G., Motevalli, M., and Hursthouse, M.B. (1986). *J. Chem. Soc. Dalton trans.*, 1141–54.

4.4 Isomer **a** has one symmetry plane through C_2 and C_4 and one through C_1 and C_3. It will have two ring carbon signals. The protons should give an AB quartet as 'up' and 'down' protons will be coupled. Isomer **b** has a centre of symmetry; it should have two ring carbon signals. All the ring protons are equivalent, and will give a singlet. The symmetry planes in isomer **c** are in the plane of the ring and through C_1 and C_3; there will be three carbon signals, but only one proton signal.

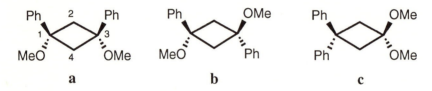

a   b   c

4.5 Yu, C., Dumoulin, C.L., and Levy, G.C. (1986). *Magn. reson. chem.*, **23**, 952–8.

4.6 Farina, M. (1987). *Topics in stereochemistry*, **17**, 31–34.

4.7 Jolly, P.W. and Mynott, R. (1981). *Adv. organomet. chem.*, **19**, 257–304.

4.8 Seddon, K.R, Turp, J.E., Constable, E.C. and Wernberg, O. (1987). *J. Chem. Soc. Dalton trans.*, 293–6.

4.9 Casy, A.F., Iorio, M.A., and Madani, A.E. (1987). *Magn. reson. chem.*, **25**, 524–30.

4.10 Pearce, C.M. and Sanders, J.K.M., (1990). *J. Chem. Soc. Perkin trans. I*, 409–1.

4.11 Aldersley, M.F., Dean, F.M., and Mann, B.E. (1986). *J. Chem. Soc. Perkin trans. I*, 2217–22.

4.12 Adapted from Lindon, J.C., Baker, D.J., Farrant, R.D., and Williams, J.M. (1986). *Biochem. j.*, **233**, 275–7.

4.13 Boyd, R.K., DeFreitas, A.S.W., Hoyle, J., McCulloch, A.W., McInnes, A.G., Rogerson, A., and Walter, J.A. (1987). *J. biol. chem.*, **262**, 12406–8.

4.14 Ziffer, H., Bax, A., Highet, R.J., and Green, B. (1988). *J. org. chem.*, **53**, 895–6.

4.15 Ichihara, A., Nonaka, M., Sakamura, S., Sato, R., and Tajimi, A. (1988). *Chemistry letters*, 27–30.

4.16 Thomas, R.D., Clarke, M.T., Jensen, R.M., Young, T.C. (1986). *Organometallics,* **5**, 1851–7.

4.17 Fontaine, X.L.R., Fowkes, H., Greenwood, N., Kennedy, J.D. and Thornton-Pett, M. (1986). *J. Chem. Soc., Dalton trans.*, 547–52.

4.18 Packett, D.L., Syed, A. and Trogler, W.C. (1988). *Organometallics*, **7**, 159–66.

5.1 Nelson, J.T. and Nelson, P.H. (1987). *Magn. reson. chem.*, **25**, 309–10.

5.2 Adapted from Pelc, B. and Sanders, J.K.M. (1972). *J. Chem. Soc. Perkin trans. I*, 1219–20.

5.3 Constable, E.C. (1985). *J. Chem. Soc., Dalton trans.*, 2687–9.

5.4 Gidley, M.J. and Sanders, J.K.M. (1982). *Biochem. j.*, **203**, 331–4.

5.5 Verhage, M., Hoogwater, D.A., Reedijk, J., and van Bekkum, H. (1979). *Tetrahedron letters*, **14**, 1267–70.

5.6 Mahaim, C., Carrupt, P.-A., Hagenbuch, J.-P., Florey, A., and Vogel, P. (1980). *Helvetica chim. acta*, **63**, 1149–57. In Table 1 of this paper, structures **9** and **11** are accidentally interchanged. Only five of the possible six isomeric products were obtained.

5.7 Dhawan, B. and Redmore, D. (1988). *J. chem. res. (S)*, 34–5.

5.8 Kazlauskas, R., Murphy, P.T., Quinn, R.J., and Wells, R.J. (1976). *Tetrahedron letters*, **11**, 4451–4.

5.9 Chamchaang, W. and Pinhas, A.R. (1988). *J. Chem. Soc. chem. commun.*, 710–1.

5.10 Takeuchi, M., Nakajima, M., Ogita, T., Inukai, M., Kodama, K., Furuya, K., Nagaki, H., and Haneishi, T. (1989). *J. antibiotics*, **42**, 198-205.

5.11 Baxter, R.L., Hanley, A.B., Chan, H.W.S., Greenwood, S.L., Abbot, E.M., McFarlane, I.J., and Milne, K. (1992). *J. Chem. Soc. Perkin trans. I.*, 2495–502.

5.12 Arai, K. and Oki, M. (1976). *Bull. Chem. Soc. Japan*, **49**, 553–8; for a review of this and related work see Oki, M. (1989). *Pure appl. chem.*, **61**, 699–708.

5.13 Compound **2**, Kimata, T., Natsume, M., Marumo, S. (1985). *Tetrahedron letters*, **26**, 2097–100.

5.14 Highet, R.J. and Batterham, T.J. (1964). *J. org. chem.*, **29**, 475–6.

5.15 Pascal, R.A., Baum, M.W., Wagner, C.K., Rodgers, L.R., and Huang, D.-S. (1986). *J. Am. Chem. Soc.*, **108**, 6477–82. For a description of how fraud in the French wine industry is detected using deuterium NMR, see Cahn, R.W. (1989), *Nature*, **338**, 708–9.

5.16 Compound **1** in Silverstein, R.M., Rodin, J.O., Wood, D.L., and Browne, L.E. (1966). *Tetrahedron*, **22**, 1929–36; spectroscopic units have been updated, and ^{13}C data have been added.

5.17 Schrobilgen, G.J. (1988). *J. Chem. Soc. chem. commun.*, 863–5.

5.18 Daloze, D., Braekman, J.C. and Pasteels, J.M. (1986). *Science*, **233**, 221–2.

5.19 The antibiotic is Anisomycin, **a**. NMR assignments are shown on structure **b**.

a   b

The aromatic region is readily built up from the appearance of signals A and B and the NOE from E, while the infrared and mass spectrum indicate the presence of an acetate group. The connectivity of the remainder of the spectrum is derived from the COSY spectrum, while the attachment point to the aromatic system is established by the NOE from H and I to A. The single DBE that remains after accounting for the acetate carbonyl group must be a ring; as one exchangeable proton is on oxygen, the ring must contain nitrogen. There are relatively few possibilities now. The 4.7 ppm signal must be attached to the acetate. The stereochemistry is established partly by the F to C NOE and partly by the very small J_{CD}.

These structural conclusions are supported by the EI mass spectral fragmentations summarized below. The CI spectrum shows MH+, M-AcOH, and the same fragmentations as the EI spectrum.

5.20 Ernst, R.R. and Meier, B.H. (1979). *J. Am. Chem. Soc.*, **101**, 6441–2.

5.21 Mueller-Westerhoff, U.T., Nazzal, A., and Prössdorf, W. (1981). *J. Am. Chem. Soc.*, **103**, 7678–81.

5.22 Ahlberg, P. and Davidsson, Ö. (1987). *J. Chem. Soc. chem. commun.*, 623–4.

5.23 Omura, S., Nakagawa, A., Aoyama, H., Hinotozawa, K., and Sano, H. (1983). *Tetrahedron letters*, **24**, 3643–6.

Index

Nuclei

There are no Index references to ^{1}H or ^{13}C because these nuclei appear on most pages.

Techniques and effects